Lori K. Garrett

Danville Area Community College

Second Edition

Get Ready

for

A&P

Benjamin Cummings

San Francisco   Boston   New York   Cape Town   Hong Kong   London   Madrid
Mexico City   Montreal   Munich   Paris   Singapore   Sydney   Tokyo   Toronto

DEDICATION

This book is for all my students, who continually educate me about life and love, and who make teaching so very enjoyable.

I dedicate this book, the now-pickled tip of my left pinkie that I sacrificed during the first edition, and more than a bit of my heart to John Hoagland, for taking care of me, believing in me, and sticking around. Finally, I want to add a note in loving memory to Danielle Nicole, who wasn't here to help me type this edition.

Editor-in-Chief: Serina Beauparlant
Development Manager: Barbara Yien
Project Editor: Sabrina Larson
Media Editor: Erik Fortier
Assistant Editor: Nicole Graziano
Managing Editor: Wendy Earl
Production Supervisor: Sharon Montooth
Cartoonist: Kevin Opstedal
Cover Artist: Chris Lochinski
Text Designer: Seventeenth Street Studios
Cover Designer: Riezebos Holzbaur Design Group
Photo Researcher: Travis Amos
Compositor: The Left Coast Group
Senior Manufacturing Buyer: Stacey Weinberger
Marketing Manager: Derek Perrigo

Learning Styles Assessment on pages 5–7
©Marcia L. Conner, www.agelesslearner.com

Benjamin Cummings
is an imprint of

www.pearsonhighered.com

ISBN 10: 0-321-55695-X; ISBN 13: 978-0-321-55695-0

3 4 5 6 7 8 9—CRS—12 11 10 09
Manufactured in the United States of America.

Contents

Welcome to the fascinating world of anatomy and physiology! Most students take A&P because they have to—it's a required part of their educational curriculum. But I hope you quickly discover how amazing the human body is and become intent on learning all that you possibly can about it. After all, you *are* one!

For many reasons, students sometimes do not succeed in their first anatomy and physiology class. If you're reading this Preface, you probably have a strong desire to succeed, and you know the competition for admission into numerous educational programs is increasing. You are likely keenly aware that you can't just pass your classes—you need to do quality work and truly master the course content. This book offers you the opportunity to enhance your performance in this rigorous course. It is designed to help you gear up to be successful.

You may be using this book before your course officially begins. Your instructor may have assigned it to you as homework during the first week or two of class. Perhaps you are using it on your own and will come back to it periodically throughout the semester as you move along in your course. However you use it, the purpose of the book remains the same. The goal is to help you get a strong start in anatomy and physiology, and to master the material not just for exams, but for your future as well. *Get Ready for A&P* contains six relatively short and interactive chapters that engage you every step of the way. You'll read, but you'll also frequently do activities.

The book starts with basic study skills in **Chapter 1**. As with any course, you will get out of A&P what you put into it, and it will take significant time and effort on your part

to succeed. This chapter helps you focus and manage your time so you can first find time to study, then use your study time effectively. After exploring different learning styles you can assess which style best fits you, then discover specific study strategies that compliment your preferred style. You'll assess your current habits as a student and learn specific tips and strategies to help you study better. Specific tips will help you write your notes, read textbooks, and take tests.

Chapter 2 covers basic math skills. Anatomy and physiology are sciences, and all science involves at least some math. This chapter takes you from basic math operations through reading and interpreting numerical information in graphs and tables—the math you'll need for a head start in your course.

Many of the words in your A&P class will sound foreign to you, and well they should! Most of the terms come from Latin or Greek. Knowing the terms underlies all aspects of learning in this class. In **Chapter 3**, Terminology, we look at how the words are built and learn some simple tricks that will rapidly expand your A&P vocabulary and have you talking like a pro!

The second half of the book gets more specific and parallels some of what you will cover in the first few chapters of your A&P textbook. In **Chapter 4**, we cover body basics: some general biological principles that guide how the body works, and a quick overview of each organ system. You'll begin to understand how your body is arranged and how the different systems function.

In **Chapter 5**, we tackle some basic chemistry. This chapter gives you the basics, from atoms to organic molecules. We discuss some neat tricks for gaining information from the Periodic Table, and see how atoms join together to form molecules. If you can understand bumper cars, you can understand bonding!

Finally, in **Chapter 6**, we explore cells. We are made of trillions of them, and almost everything that happens in our bodies occurs inside our cells. We discuss basic cell structure, the cell life cycle, and cell reproduction.

Now that you know the roadmap for this book, let's explore the stops you'll find along the way. Here are the special features in each chapter, designed to keep you involved and to make you a better student in A&P:

- *Your Starting Point* tests your grasp of the chapter content before you start. Answers are provided for all of these except in Chapter 1, where the answers are personal.

- *Quick Check* asks you to recall or apply what you just read, to keep your eyes from scanning the page while your brain is on vacation. Answers are provided on the same page.

- *Picture This* asks you to visualize scenarios and then answer questions about them to help you better understand the topics.

- *Time to Try* is a simple experiment or quick assessment in which you perform an active exercise to provide hands-on learning.

- *Why Should I Care?* highlights the relevance of the material so you understand its importance in the big picture.

- *Reality Check* assesses whether you really "got" the material.

- *Keys* highlight main themes or statements for reinforcement and easy review.

- *Running Words* list key terms from the chapter to help you start your own running vocabulary list by writing each term in a notebook, then defining it.

- *What Did You Learn?* end-of-chapter quizzes may include short answer, multiple choice, or matching exercises. The answers appear at the end of the book.

Finally, check out the online companion site for *Get Ready for A & P,* where you can study and quiz yourself with interactive tutorials, quizzes, animations, flashcards, and an audio glossary.

I wrote *Get Ready for A&P* in a conversational tone because that is how I teach and because science is too often turned boring by boring presentation. Science shouldn't be stuffy—it should be fun! Learning should be peppered with giggles, salted with silliness, and dotted with *AhHA!* moments. Sometimes it seems the terminology alone can put you in a trance, so why should the style?

Now it's time to dig in. So get comfortable, and *Get Ready for A&P!*

Preface for Instructors

For some time, we've all grappled with the same problem: too many students entering our classrooms fail, often through no fault of their own. At every conference I attend, instructors lament the low A&P retention rates. As teachers, we want all of our students to succeed—we hope our work has value and that their successes reflect well on us. As my colleague puts it, they may be our future health care providers—I hope they learned *something*! With retention numbers under increasing scrutiny on campuses, instructors must look harder at what issues hinder student success and then address them. Several factors contribute to the low success rates in A&P, and *Get Ready for A&P* takes many of these into account. I wrote this book because an increasing number of my students were entering my A&P course with limited study skills and insufficient background in science and math. My goal was to write a book that could help students get up to speed in these areas. Whether your students are deficient in study skills or academic background, *Get Ready for A&P* is their solution.

If you read the *Preface for Students,* you know the basics about the book. It's written in a friendly, conversational style to help students feel comfortable with science. It's also rich in pedagogical features that cater to most learning styles and keep students actively engaged. Each of *Get Ready for A&P*'s six chapters starts at a very basic level and builds the knowledge base from there, bringing all of your students to common ground. For many students, **Chapter 1—Study Skills**—is the most important. They'll learn how to be active, engaged learners, to use resources efficiently, and to be accountable for their performance. For the Second Edition, I expanded Chapter 1 to include more about learning styles, goal-setting, active reading, and concept maps, for example; new topics include sleep deprivation and memory, and minimizing stress and test anxiety.

Not all students are deficient in math, but most will benefit from the quick review offered in **Chapter 2 (Basic Math),** which now includes calculating grades and examples of good and poor graph design. **Chapter 3 (Terminology)** teaches students how to decipher those long terms we throw at them and stresses the importance of understanding the language. They'll get plenty of practice using roots, prefixes, and suffixes to define and build terms. **Chapter 4 (Body Basics)** covers universal biological principles governing physiology and now also includes positive feedback.

In response to reviewers' requests, most additions are in **Chapter 5 (Chemistry)** and **Chapter 6 (Cells),** which cover content students should know before coming to us. New to Chapter 5 are electrolytes and pH, monomers and polymers, and dehydration synthesis and hydrolysis; the discussion of basic organic compounds has expanded. Chapter 6 now includes peroxisomes, hypo- and hypertonic solutions, DNA replication, and an overview of meiosis.

The ***Get Ready for A&P* website** offers a diagnostic test and a follow-up test, chapter pre- and post-tests, animations, tutorials, activities, and useful web resources to further student learning in their preferred environment—online. We've updated the Second Edition website to include more tutorials and activities.

Finally, we've **expanded the online assessment options** for the Second Edition, providing you with two versions of the diagnostic test, the cumulative test, and the chapter pre- and post-tests to use in your course management system, allowing you to assess your students' progress.

But, you may ask, **"How can I incorporate this book into my curriculum?"** I know your semester is packed, but the advantages this book provides your students make it well worth adding. The book's small size and interactive format make it easy for your students to work through *Get Ready* in very little time. Many instructors prefer that their students finish *Get Ready* before starting A&P. If your school has a pre-A&P course, *Get Ready* is a perfect companion for it. Is there a prerequisite to your A&P class? Make *Get Ready* part of that course as a platform for launching more in-depth lectures or assignments outside of class. Other instructors suggest *Get Ready* workshops before students enter A&P or at the beginning of a semester—perhaps requiring it for students who've been out of school awhile or who have weak science or academic backgrounds.

Many instructors use *Get Ready* during the A&P course. Some instructors simply have the book available in the bookstore, while others list it as "recommended" or "optional" for the A&P class. Students can work through it on their own time and turn in the end-of-chapter quizzes. You might give them the first week or two to complete the book. Some instructors, for example, assign the assessments or the diagnostic test on the companion website either for homework or as extra credit. Alternatively, consider using *Get Ready* throughout the semester. I recommend Chapters 1 and 2 for the first week, then Chapters 3 and 4 during the second week—their content parallels the first chapter of most A&P textbooks. If you cover basic chemistry, assign Chapter 5 of *Get Ready* for preparation, and then assign Chapter 6 when you discuss cell structure and function in class. You can also make assignments from *Get Ready's* companion website.

Because many of today's students prefer online courses, and to keep your life simple, we've also launched a **Get Ready for A&P Online**

Course. The ten lessons include reading assignments from the book, lecture screens, exercises, online discussions, and gradable assignments. Each lesson is highly interactive and engaging for today's students and includes "Check Your Understanding" screens for self-assessment through activities ranging from bowling and crossword puzzles to more traditional matching and multiple-choice questions. *Get Ready Online* is fully customizable, making it ideal for *Get Ready* workshops or pre-A&P courses. Use the online course as is or modify it to meet your needs—add your own material or remove what you don't want. It can easily be blended into a preexisting course, too. This online course was developed from carefully selected lesson and course objectives aligned with Bloom's taxonomy to ensure that your students benefit greatly while having fun. For more information about the *Get Ready for A&P Online* course, visit http://www.pearsonls.com/getready.

Ancillary materials for A&P abound—and many are excellent—but almost none address our most pressing problem: student deficiencies. We assume they can study, but in reality many of our students neither learned how to study nor were ever required to do so. We know they come with some science background from high school, yet they often recall little. Too many of our students struggle daily, yet make little progress. Our frustration pales compared to theirs. For many of them, failing A&P can end their hopes of a future career by blocking their admission into a desired program or by instilling the belief that, for them, a college degree is unobtainable. And for us-how nice it would be to begin class with our students all at a common starting point! I wrote *Get Ready for A&P* to give all of our students an equal chance from the beginning and a fair shot at a promising future.

Lori Garrett

Acknowledgments

What do you mean, Second Edition? I remember thinking that when my good pal, Serina Beauparlant, told me it was time to get back to work on this. It didn't seem very long ago that I wrote the first one. And when I thought about the positive responses we've had to the first edition, part of me said *"Don't mess with it—you'll screw it up!"* but off we went. I have to thank all the first edition team—Claire Alexander, Jessica Brunner, Travis Amos, Randall Goodall, and Seventeenth Street Studio, without whom there would have been no need for a second edition.

This time around, some veterans returned to help keep me in line. I again thank Wendy Earl for her excellent management (I can be a tad unruly), copyediting, and gentle guidance with ever encroaching deadlines. Wendy, thanks for the long-distance hand-holding near the end. I again thank Kevin Opstedal for drawing more great cartoons to keep this fun for us all. And I thank Erik Fortier for his hard work on the companion website through its many renditions to date—with more to come . . . *I know, I know. . .*

I also had the pleasure of working with Sabrina Larson. Sabrina, any woman who eats beignets heaped with powdered sugar at 7:30 AM at Café du Monde while wearing a dark business suit and remains poised and tidy through the entire ordeal earns a special place in my heart! Thanks for trying to make this all easy for me. I also thank Sharon Montooth for her hard work overseeing the production phase, especially when we were trying to squeeze too much within the covers. I am eternally thankful for Nicole Graziano who is always there at Benjamin Cummings, calmly holding the place together behind the scenes. Thanks, too, to Frank, who keeps all of our encounters entertaining and memorable. I also thank the Pearson Course Connect Team—David Daniels, Erik Styles, and Dawn Stapert—for helping me develop the online *Get Ready for A&P* course and opening new doors for our students.

With the greatest respect, I thank Serina Beauparlant for keeping me off the streets—sometimes I forget which job is my full-time day gig! Serina, you make me feel invaluable and have been remarkably accommodating. Your dedication is amazing—who else would travel to Illinois on the most brutal day of winter when all businesses, schools, and many roads are closed to visit an author? You go above and beyond, not just for me, but for all of your authors, and your dedication to improving education is inspiring. I'm thrilled to have you as a business partner, but even more so as a very dear friend. Whether we're talking politics or production deadlines, I always enjoy it.

I thank all of the students and instructors whom I've had the pleasure of meeting either online or at conferences and workshops since the first edition came out, who have reaffirmed for me the need and value of this project. I thank the excellent reviewers of the first edition for their positive responses and for providing keen insight and advice on ways I could make *Get Ready* even more than it already was. Finally, I again acknowledge my family at Danville Area Community College. I thank my students for doing their jobs well so I look good in mine. I thank Dean Janet Redenbaugh for her constant support of all my undertakings and for keeping our division the happiest place on campus, and I thank John Hoagland for putting up with all the insanity and chaos in my life (although, in fairness, he creates much of it!)

-Lori

REVIEWERS

Kim Anthony Aaronson
Truman College, Harold Washington College,
Columbia College

Kathleen Azevedo
Las Positas College

Chester P. Cooper
Odessa College

Jose L. Fierro
Florida Community College at Jacksonville

Anthony Gaudin
Ivy Tech Community College—Columbus

Chaya Gopalan
St. Louis Community College at Florissant
Valley

Judy Harris
Kennebec Valley Community College

Jacqueline Homan
South Plains College

Jody Johnson
Arapahoe Community College

Stephen Lebsack
Linn Benton Community College

Karen M. MacKenzie
Greenville Technical College

Ken Malachowsky
Florence-Darlington Technical College

William McCracken
Tallahassee Community College

Claire Miller
Community College of Denver

Kerry Smith
Oakland Community College

Mary Weis
Collin County Community College

1 Study Skills

The Proper Care and Feeding of a Human Brain

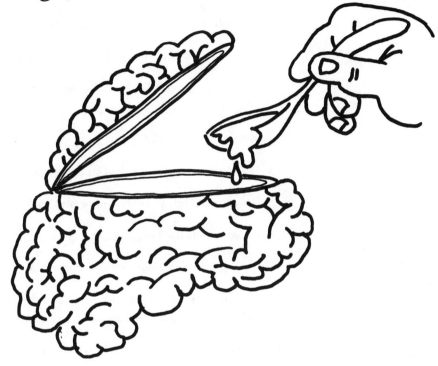

When you complete this chapter, you should be able to:

■ Understand your preferred learning style and study strategies that emphasize it.

■ Utilize skills that will help you get the most benefit from lectures, labs, and readings.

■ Draft a written schedule that includes adequate study time.

■ Know how to prepare well for an exam.

■ Take responsibility, realizing that you are ultimately accountable for your own success or failure.

Your Starting Point

Answer the following questions to assess your study habits.

1. How often do you read a course textbook? _____

2. How many days of the week do you study for one course? _____

3. Do you study hard the day before an exam, but rarely between exams?

   _____

4. Where do you study? _____

5. How long should you spend studying outside of class? _____

6. Do you schedule your study time and stick to it? _____

7. Do you study hard or hardly study? _____

8. Do you mostly memorize when studying for a test? _____

9. Do you have a good support group of family and friends who encourage you? _____

10. Do you quiz yourself when studying? _____

Welcome to the exciting and sometimes challenging world of anatomy and physiology! You will quickly discover how amazing the human machine truly is—a curious marvel of complexity that is simultaneously surprisingly simple. I hope you will be fascinated by learning how your own body is built (anatomy) and how it works (physiology). Interest in your subject matter always makes it much easier to learn.

Still, no matter how exciting your anatomical explorations may be, your course may, at times, seem rigorous and demanding. You've taken a great first step by turning to this book to jump-start your studies. This book is meant to help you enter the course with a well-planned strategy

Answers: Answers will be individualized, except for #5—you should spend 2–3 hours studying for each hour of class time.

for success and with confidence in your basic science knowledge. The purpose of this chapter is to help you "train your brain" to make your learning process easier and more efficient.

Why Should I Study Anatomy and Physiology?

Most students take anatomy and physiology because it is required for their educational programs. Sometimes when something is required, we do it only because we have to without considering what benefits the task might hold for us. Unfortunately, some students use that approach for anatomy and physiology. Certainly it is easier to study something if you understand why it matters, and this course is no exception.

PICTURE THIS

Until recently your car has run perfectly, but now the engine occasionally quits running and is difficult to restart. Assuming you have little knowledge of auto mechanics, you are not likely to solve the mystery or make repairs yourself. You take it to an auto mechanic, who will consider how your car is malfunctioning—its symptoms, if you will—and then fix it. What knowledge will the mechanic need to accomplish that goal? _____

_____

_____

In what ways are people in health- and medical-related fields similar to the auto mechanic? _____

_____

Why do they need to fully understand anatomy and physiology?

_____

_____

Now consider your own future—what is your planned career?

_____

Why will you need to know anatomy and physiology?

_____

_____

Many anatomy and physiology students plan careers in a medical or health field. Others may be heading into kinesiology, athletic or personal training, perhaps biomechanics or bioengineering, and many other fields. These career areas share a common thread—anatomy and physiology form the foundation on which they are all built. Now, back to our example. To understand your malfunctioning car, the mechanic must first fully understand the parts of your car—how they fit together and how they normally function, just as you will need to understand the parts of the human body and their normal functions. Finally, there is a simpler reason why you should care about learning anatomy and physiology. The human body is an amazing machine, and you own one. Anatomy and physiology are your owner's manual.

To Thine Own Self Be True: **Learning Styles**

What *is* the best way to learn these subjects? A tremendous amount of research has explored how people learn, and there are many opinions. One common and simple approach considers which of the senses a learner relies on the most—sight, sound, or touch:

- Visual learners learn best by *seeing*.

- Auditory learners learn best by *hearing*.

- Tactile (kinesthetic) learners learn best by *doing*.

TIME TO TRY

Let's uncover your learning style. Look at **Table 1.1.**

1. Read an activity in the first column, then read each of the three responses to the right of that activity. Mark the response that seems most characteristic of you.

2. Do this for each row. Then add the marks in each column and write the total in the bottom row.

3. You will likely have a higher total in one column. That's your primary learning style. The second highest number is your secondary style.

   My primary learning style is: _____

   My secondary learning style is: _____

TABLE 1.1 Assessing your learning style.			
Activity	**Column 1**	**Column 2**	**Column 3**
1. While I try to **concentrate** . . .	I grow distracted by clutter or movement, and I notice things in my visual field that other people don't.	I get distracted by sounds, and I prefer to control the amount and type of noise around me.	I become distracted by commotion, and I tend to retreat inside myself.
2. While I am **visualizing** . . .	I see vivid, detailed pictures in my thoughts.	I think in voices and sounds.	I see images in my thoughts that involve movement.
3. When I **talk to someone** . . .	I dislike listening for very long.	I enjoy listening, or I may get impatient to talk.	I gesture and use expressive movements.
4. When I **contact people** . . .	I prefer face-to-face meetings.	I prefer speaking by telephone for intense conversations.	I prefer to interact while walking or participating in some activity.
5. When I **see an acquaintance** . . .	I tend to forget names but usually remember faces, and I can usually remember where we met.	I tend to remember people's names and can usually remember what we discussed.	I tend to remember what we did together and may almost "feel" our time together.
6. When I am **relaxing** . . .	I prefer to watch TV, see a play, or go to a movie.	I prefer to listen to the radio, play music, read, or talk with a friend.	I prefer to play sports, make crafts, or build something with my hands.
7. While I am **reading** . . .	I like descriptive scenes and may pause to imagine the action.	I enjoy the dialogue most and can "hear" the characters talking.	I prefer action stories, but I rarely read for pleasure.

▶

TABLE 1.1 Assessing your learning style, continued.

Activity	Column 1	Column 2	Column 3
8. When I am **spelling** . . .	I try to see the word in my mind or imagine what it would look like on paper.	I sound out the word, sometimes aloud, and tend to recall rules about letter order.	I get a feel for the word by writing it out or pretending to type it.
9. When I **do something new** . . .	I seek out demonstrations, pictures or diagrams.	I like verbal and written instructions, and talking it over with someone else.	I prefer to jump right in to try it, and I will keep trying and try different ways.
10. When I **assemble something** . . .	I look at the picture first and then, maybe, read the directions.	I like to read the directions, or I talk aloud as I work.	I usually ignore the directions and figure it out as I go along.
11. When I am **interpreting someone's mood** . . .	I mostly look at his or her facial expressions.	I listen to the tone of the voice.	I watch body language.
12. When I **teach others how to do something** . . .	I prefer to show them how to do it.	I prefer to tell them or write out how to do it.	I demonstrate how it is done and ask them to try.
TOTAL:	**Visual:** _____	**Auditory:** _____	**Tactile/Kinesthetic:** _____

[Source: Courtesy of Marcia L. Conner, www.agelesslearner.com]

TABLE 1.2   The three learning styles and helpful techniques to use in your studies.

 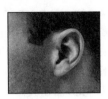 

	Visual	**Auditory**	**Tactile**
Techniques to use	❏ Sit close to the teacher.	❏ Listen carefully to your teacher's voice.	❏ Highlight while reading.
	❏ Take detailed notes.		❏ Write your own notes in class and while reading.
	❏ Draw pictures.	❏ Read the textbook and your notes out loud.	
	❏ Make flow charts.	❏ Tape record lectures and listen to them later.	❏ Transfer your notes to another tablet or type into your computer.
	❏ Use flash cards.		
	❏ Focus on the figures, tables, and their captions.	❏ Listen during class instead of writing notes.	❏ Doodle and draw as you read.
			❏ Build models.
		❏ Work in a study group.	❏ Create and conduct your own experiments.
	❏ Try coloring books and picture atlases.	❏ Discuss the material with others.	❏ Walk or stand to read.
	❏ Use visualization.		❏ Use A&P coloring books.
			❏ Use flash cards.

Now that you know your primary and secondary learning styles, you can design your study approach accordingly, emphasizing activities that use your preferred senses. Look closely at your scores, though. If two scores are rather close, you already use two learning styles well and will benefit from using both of them when studying. If your high score is much higher than your other scores, you have a strong preference and should particularly emphasize that style. Most people use a combination of learning styles.

In addition, information coming in through different senses reaches different parts of your brain, activating more neural pathways that allow you to learn. The more of your brain that is engaged in the learning process, the more effective your learning will be, so try strategies for all three styles and merely emphasize your preferred style over the others. You'll know which strategies work best for you. We'll consider some strategies that you might try for each style; these ideas are summarized for you in **Table 1.2**.

VISUAL LEARNERS

If you are a visual learner, you rely heavily on visual cues. You notice your teacher's mannerisms, expressions, gestures, and body language. Seeing these cues is especially helpful, so sit at the front of the classroom, close to the teacher. You tend to think in pictures and learn well from visual aids such as diagrams, illustrations, tables, videos, and hand-outs. Here are some strategies for you.

- In class, take detailed notes and make sketches.

- When studying on your own, draw pictures that relate to the information, make flow charts and concept maps, use flash cards, focus on illustrations and tables in your textbook, and read the captions that accompany them.

- Use anatomy and physiology coloring books and picture atlases.

- Mentally visualize the material you are studying and imagine yourself acting out processes. For example, to learn major blood vessels, you might imagine yourself swimming through them.

AUDITORY LEARNERS

If you are an **auditory learner**, you learn well from traditional lectures and discussion. You listen carefully to your teacher's vocal pitch, tone, speed, and mannerisms. Material that you struggle with while reading becomes clearer when you hear it. Here are some strategies for you.

- Read the textbook and your notes out loud.

- Tape record the lectures so you can listen to them later. Taping lectures also allows you to listen during class instead of focusing on writing, which is less beneficial for you.

- Work in a study group, and discuss material with your teacher, lab group, and friends.

TACTILE LEARNERS

If you are a **tactile learner**, you learn best by actively participating and doing hands-on activities. You may become bored easily in class from sitting still too long and start fidgeting or doodling. You need to do something physical while studying and learning. Here are some ideas for you.

- Try using a marker to highlight important information while you are reading.

- Write out your own notes in class and while reading the textbook. Later, transfer your notes to another tablet or type them into your computer.

- Draw pictures of appropriate material as you read.

- Build models of anatomical structures using clay or other materials.

- Create and conduct your own experiments.

- Hold your book and walk while reading.

- Use anatomy coloring books.

- Make and use your own flash cards.

- Keep your hands and your mind busy together.

We've considered only one set of learning styles, but there are many other systems. The VARK system, for example, recognizes four styles: visual, aural, read/write, and kinesthetic. Another system, called Mem-letics ("*mem*ory ath*letics*") identifies seven learning styles. Many other systems, such as the Kolb Learning Style Inventory and the Myers-Briggs Type Indicator are also available, some for free and online. To gain more insight into your preferences, you can explore other systems by doing a web search for "learning styles." Whether or not you seek additional information, it's good to at least have a general idea of your learning preferences. Matching your study approach to your learning preference can help you gain the most from your study sessions. Find out what works for you.

WHY SHOULD I CARE?

Understanding your own learning style allows you to develop more effective and efficient study techniques that take advantage of your sensory preferences. By emphasizing your preferred learning style, the material will be easier to learn and will stay with you longer.

✔ **QUICK CHECK**

Homemade flash cards would be most beneficial to which two learning styles? _____ and _____.
How could they be used to benefit a learner of the third style?

_____

Putting on Your Best Face: **Getting Ready**

Many students mistakenly wait until the first lecture to start thinking about class. The key to starting your semester well is to be organized and ready when you enter the classroom. This takes advance planning, but the time invested will save you tremendous time while the semester is underway.

PUTTING IT IN WRITING

As the semester begins—preferably before—you should get organized, and that begins with making a commitment to yourself. Too often we begin a project without setting goals in advance. If you set a goal, you enter with a purpose and a direction. If you do not set a goal, it's too easy to just go along and see where you end. Take time to think about your goals for the semester.

Your goals should be **SMART**, which stands for the characteristics to incorporate into your goals:

- *Specific.* Instead of trying to "do well," perhaps strive for a specific grade.

- *Measurable.* How will you know if you achieve your goals? For example, a goal of studying for two hours a day is measurable.

- *Accountable and attainable.* Set goals for yourself that you can achieve through your own effort and accountability. Start your goals with "I will...," and be sure they are attainable. Having the top grade in class might not be possible, but you can get an "A."

Answer: They would benefit visual and tactile learners. Reading them out loud would benefit auditory learners.

■ *Realistic.* If you are a single mother with youngsters at home, it may not be realistic to set a goal of studying two hours each evening when your children need your attention.

■ *Time-based.* Set a time frame for achieving your goals and be realistic about how much time to allow. Let's say you plan to enter a nursing program. If you must take several prerequisite courses before admission, you'll not likely get in after a year if you attend school part time.

Take time to set appropriate goals for yourself. Many students underestimate how important these goals are. They will help you stay motivated by keeping you focused on why you are in school and where you are going in your life.

After you've decided on your goals, write them down to give them more importance. Once you've written them, be firmly committed to them. To reinforce these goals, write them on an index card or type them into your computer, print the sheet, and place it in a prominent location in your study area so you'll see your goals every day.

TIME TO TRY

Set three main goals for yourself that relate to this class, and write them below. Explain why achieving each goal is important to you.

Goal 1: _____

*It is important to me because:*_____

_____

Goal 2: _____

*It is important to me because:*_____

_____

Goal 3: _____

*It is important to me because:*_____

_____

PULLING IT ALL TOGETHER

I'm amazed when students show up for a test with no writing utensils! Don't let that happen to you. The more organized you are, the more efficient you will be, so let's organize what you will need for class. Categorize the items by what you take to class every day, what remains at home in your study spot, and optional items that are nice, but nonessential, additions. Use the checklist provided for you in **Table 1.3**. Search your house and you'll likely find that you have many of these items already. Most items can be bought at your college bookstore, but many are available at discount stores. We will discuss some of these items specifically.

You'll be going back and forth to class a lot, so it is most efficient to keep all the items you might need for class in one place. To haul them, most students use a book bag, backpack, or briefcase. An advantage to using one of these is that you can load it up with the essentials so that they are always ready to walk out the door with you. Let's discuss some of the items to pack.

You need a pocket-sized day planner that has plenty of room for writing and that you can keep with you at all times. Or you may opt for a personal organizer portfolio or an electronic organizer. Select one you like, because you'll use it every day. In it, write all important dates you already know—when classes begin, holidays, last day to withdraw from a class, when finals begin. Enter all class times, your work schedule, and any other known time commitments. Try to keep your day planner current so you always know how your time is being spent and can plan ahead.

If it's not part of your day planner, you need a separate To Do List. You will write all assignments and due dates on this list. You want one single To Do List for all of your classes as well as non-school activities, because they must all be done from the same pool of time. Writing them down allows you to view the entire list and review the deadlines for each item so you can easily prioritize, doing the assignments in the order in which they are due.

Maintain a record of all grades you receive (**Figure 1.1**). For each graded item, list what it is, when you turned it in, when you got it back, how many points you received, how many points were possible, and any additional notes. Once you know how your grade will be determined for

TABLE 1.3 Organizer's checklist.

Item	✎✗
To take to class each day:	
Book bag/backpack/rolling carrier	
Textbook/lab manual	
Pocket-sized day planner	
To Do List	
Separate notebooks for each course	
Copy of class schedule with buildings and room numbers	
Several blue or black ink pens	
Several #2 pencils	
Small pencil sharpener	
2–3 colored highlighter pens	
Small stapler	
Grade record sheet for each course	
Calculator	
At home:	
Master calendar	
Separate file or folder for each course	
Loose notebook paper	
Index cards for making flash cards	
Computer paper (if I have a computer)	
More writing utensils (pens and pencils)	
Stapler	
Calculator	
Scissors	
Paper clips	
Optional:	
Personal organizer	
Anatomy and Physiology coloring book(s)	
Colored markers/pencils	
Small tape recorder to record lectures/readings	
Recording tapes	
Extra batteries	
Anatomy atlas	
Medical dictionary	

Graded item	Date turned in	Date returned	My score	Possible points	Notes
Lab 1	9/6	9/13	10	10	Worked with Emily, Mike, Tom
Quiz 1	9/7	9/9	18	20	Study terms again
Lab 2	9/13	9/16	6	10	Messed up the math!
Pop Quiz	9/14	9/16	5	5	From yesterday's lecture.
					I was ready!
Quiz 2	9/21	9/25	19	20	Forgot to answer one question!

FIGURE 1.1   A sample grade record for keeping track of your progress.

the course, you can use this to keep track as you go along. It also provides a backup in case there is any confusion later about your grade or your work. You should also maintain a separate folder for all graded items that are returned to you.

Check with your instructor to see which books you should bring to class. Typically, you may not need your textbook in lecture, but you may need it in lab. If your course uses a lab manual, always take it to lab, but you may not need it in lecture.

And for both lecture and lab, always carry the basics. You need a notebook for note-taking. If you come to my class without a writing utensil, it says you do not think that what I am saying is important enough for notes. If you ask me for a stapler before turning in an assignment, it says you threw it together with little thought. Pens, pencils, erasers, staples, paper, highlighters, colored pencils, index cards, paper clips—these are just some items you may find useful. Replenish your supply as needed.

Set up your home study space like a home office. Be sure to have all the essential office supplies on hand—plenty of writing utensils, paper, a stapler, and so on. A critical part of the home study area is the master calendar. There are large desktop versions and wall charts, for example.

You could use a calendar feature on your computer, but the more visible the calendar is, the more often you will look at it. This calendar should be large enough to accommodate plenty of writing, so think BIG! Each day, you should add anything that you put in your day planner or on your To Do List to this master calendar. All time commitments should be entered, so also add all personal appointments and vacations. This is how you will schedule your life while in school, and the practice will likely stick with you far beyond that.

You may have already started your class before using this book. If so, you certainly cannot do all these things before class at this point, but it is never too late to get organized. So, go get it together!

✔ **QUICK CHECK**

To be successful in class, your effort should start before class begins. What are some tasks you should do before the first day of class?

_____

_____

_____

I Hate to **Lecture** on this, but Can You Hear Me Now?

Welcome to class! Imagine that it is the first day. You walk into class.

Where do you sit? _____

Why do you sit there? _____

The best seat in the house is front and center. Obviously not everyone can sit there, but you should arrive early enough to sit within the first few rows and as near to the middle as possible. You want an unobstructed view of the instructor and anything he or she might show, because anatomy is often a very visual course. People sitting on the sides or in the back often do not want to be called on, or they want to be in

Answers: Set and write down your goals, organize the items you will need for class and at home, pack your carrier, start your day planner and master calendar, and organize your study space.

their own space. They often are not very engaged in the class. Don't let that be you. To succeed, you need to focus all of your attention on your instructor, minimize distractions, and actively participate. Instructors tend to teach to the middle of the room. In fact, if your instructor is right-handed and uses equipment, such as an overhead projector, that is positioned on the right, the instructor's focus shifts to his or her right. You want to see your instructor and you want your instructor to see that you are present, actively listening, and engaged.

Some instructors provide lecture notes so you can sit back and really think about what is being said. Notes or not, you need to get all the information you can from each lecture. Remember your learning style and use techniques that enhance it. We will discuss note-taking momentarily, but consider tape recording the lectures. That way you miss nothing, and you can listen to the tape repeatedly, rewinding as needed. Another good technique is to write out your own notes while listening to the tape, then listen again while reading your notes and making corrections. This combination strongly reinforces the material.

Always try to preview the material that will be covered before going to class. This is as simple as lightly reading the corresponding sections in the textbook. You may not understand all that you read, but it will sound familiar and be easier to comprehend as your instructor covers it in class. This preview also helps you identify new vocabulary words.

While your instructor is lecturing, don't hesitate to raise your hand to ask a question or get clarification. Many students are shy and reluctant to speak in class—you may be doing them a favor! Avoid discussing personal issues in front of the whole class—that is better done alone with the instructor, outside the classroom.

Note your instructor's gestures, facial expressions, and voice tone for clues about what your instructor finds most important. That material is likely to show up on a quiz or test. Write down any material that is particularly emphasized, or mark it in your notes. Listen carefully for assignments and write them down immediately on your To Do List. If you are not clear about the expectations of the assignment or when it is due, seek immediate clarification. Don't ever try to second-guess your instructor's expectations.

✔ QUICK CHECK

Why is it best to sit front and center in class?

_____

_____

Passing **Notes**

Anybody can take notes in class, but will the notes be good enough to help them succeed in the class? There are many strategies and models for how to take notes, and none of them is necessarily the best. Find what works for you, then use it consistently. Let's review one easy-to-use system (**Figure 1.2**).

Start with a full-sized (8.5" x 11") notebook that you will use just for this class. Take your notes on only the front side of the paper and leave about a 2" margin on the left. The margin will be used for marking key words and concepts later. At the beginning of class, date the top of the page so you know when the material was covered. During lecture, use an outline format to get as much information down as you can. Use the main concept as a major heading then, under it, indent the information discussed on that topic. When that section ends, either draw a horizontal line to mark its end or leave a couple of blank lines before you write the next major heading. Don't try to write every word—just the main ideas—and put them in your own words. Instead of writing out every example, give a brief summary or a one- or two-word reminder. Use abbreviations when possible, and develop your own shorthand. You can often drop most of the vowels in a word and still be able to sound it out later when reading it. Write legibly or your efforts will be useless later. Underline new or stressed terms and place a star or an arrow by anything that is emphasized. Be as thorough as you can, but you will need to write very quickly. The instructor will not wait for you to catch up, so both speed and accuracy are essential.

Answer: You will be more engaged in the class, have the best view and fewer distractions, and be within your instructor's focal area.

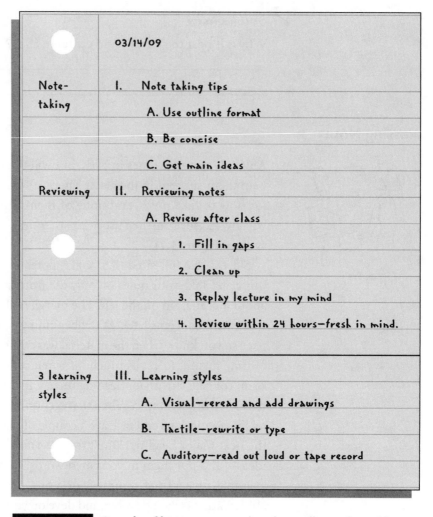

03/14/09

Note-taking	I. Note taking tips
	A. Use outline format
	B. Be concise
	C. Get main ideas
Reviewing	II. Reviewing notes
	A. Review after class
	1. Fill in gaps
	2. Clean up
	3. Replay lecture in my mind
	4. Review within 24 hours—fresh in mind.
3 learning styles	III. Learning styles
	A. Visual—reread and add drawings
	B. Tactile—rewrite or type
	C. Auditory—read out loud or tape record

FIGURE 1.2   Sample of lecture notes using the outline style and leaving room in the left margin.

As soon as possible after class, read your notes and improve them as necessary. Add anything that's missing. Make them clearer and cleaner. Put the concepts in your own words. Next, use the left margin to summarize each section—the main concept, subtopics, and key terms. The latter column will be your "Recall" column. Once you are sure all of the key ideas are in the left column, you can cover the right side of the page—the meat of your notes—and quiz yourself on the main points listed in the left column. It makes an easy way to review.

But you are not finished—if you are a tactile learner, rewrite your notes in another notebook or type them into your computer. Visual learners might type and reorganize the notes. Auditory learners can read the notes out loud or record them. You can make flash cards from the key points and terms by writing the term on one side of an index card and its definition or use on the other side. You can add drawings. Review your notes as much as you can during the next 24 hours while the lecture is still fresh in your mind.

Go through your notes several times and think about related questions that might be asked on tests. Find creative ways to quiz yourself on what was covered.

As mentioned earlier, your instructor may provide lecture notes for the class. This allows you to listen carefully and think about what the instructor is saying rather than focusing on writing. Even so, always be ready to take additional notes, noting what your teacher emphasizes or adds that's not already in the lecture notes. And you can and certainly should continue to employ the study techniques we've discussed here even if you did not take the original notes.

TIME TO TRY

Look at the sample notes in Figure 1.2. Now practice: Take notes using this style while listening to a 1-hour TV show. Capture the conversations and action in words. You can't get every word down, so paraphrase—put it in your own words so the meaning still comes across. When you're finished, assess how you did.

Can you tell who was talking? _____

Do your notes make sense? _____

Did you capture the main ideas? _____

Did you keep up or fall behind? _____

Do you have breaks to separate main conversations and action?

_____

What can you do better while taking notes in class?

_____

✔ **QUICK CHECK**

What should you do with your notes after class? _____

_____

_____

Looking for Some-body Special **In the Lab**

Your anatomy and physiology class comes in two parts: lecture and lab. Many students put most of their effort into the lecture material and disregard the lab component. Avoid this. Lab is the hands-on part of the course, and most people learn better by seeing demonstrations and actually doing the work themselves. Also, part of your grade comes from your performance in lab. Always go to lab prepared to take notes, equipped with your lab manual, if required, and your textbook if it will be needed.

When in lab, you may work with a lab partner or group. You will be expected to contribute equally to the team effort, so it is important that you arrive prepared for lab. If you have a prelab assignment, complete it before you go to class. If you know in advance what the lab will be, read through it and think about what you will be doing. Pay attention to the instructions and especially note any safety precautions. At times you will be working with very expensive equipment and specimens, and perhaps potentially dangerous materials, so always use great care.

Some students try to take shortcuts in lab so they can leave a bit early. You should value lab as a time to further explore the material covered in lectures. You get to hold the bones, microscopically examine various tissues, use models and charts, test physiological processes, and perhaps see real organs or cadavers. It is a unique aspect of your education that reinforces everything else that you are learning. You will spend much time learning structures, and the more you go over them, the better you will recall them. The more time you spend in lab, the more you learn. Remember that lecture and lab are both part of the same class and try to see how they fit together. Never leave lab early—there's always more to learn.

Answer: Review them within 24 hours, fill in anything missing, clean them up, put them into your own words, add key concepts and terms to the Recall column, add drawings, make flash cards, record them, rewrite or type them.

✔ **QUICK CHECK**

What are some of the learning advantages gained from attending lab sessions? _____

_____

 The more time you spend in lab, the better you will learn. ■

Your Secret Life **Outside of Class**

REALITY CHECK

Answer True or False to each of the following statements:

1. I study the day before a test but rarely study on a daily basis. T F

2. I mostly review my notes and don't read the textbook. T F

3. I am too busy to study each day. T F

4. When I finally get around to it, I study hard for a long time. T F

5. I get by fine with cramming. T F

JUST FOR FUN

Let's see how good a studier you REALLY are! Take a few moments to learn these terms. We will come back to this exercise a bit later.

1. **Frizzled greep.** This is a member of the *Teroplicanis domesticus* family with girdish jugwumps and white frizzles.

2. **Gleendoggled frinlap.** This is a relatively large fernmeiker blib found only in sproingy sugnipers.

3. **Borky-globed dungwinger.** This groobler has gallerific phroonts and is the size of a pygmy wernocked frit.

Stay tuned!

Answer: Lab is time for exploration, hands-on learning, collaboration, and discussion.

You made it through lecture or lab, and are ready to head home. Finally! School is done for the day, right? Not if you plan to be successful! The real work begins after class, because most of your learning occurs outside the classroom on your own. This is often the hardest part, for many reasons. We schedule many activities and set aside time for them, but studying tends to get crammed into the cracks. Too often, studying becomes what you do when you "get around to it." It is an obligation that often gets crowded out by other daily activities, and studying is often the first item dropped from the To Do List.

Too many students only study when they have to—before a quiz or exam. A successful student studies every day. The goal is to learn the material as you go rather than frantically try to memorize a large amount at the last minute. Here is something you need to know and really take to heart.

You should study for at least two to three hours for every hour spent in class. ■

Simple math shows you that if you have three lectures on Monday, for example, you should plan to spend from six to nine hours studying before the next lectures. YIKES!

SCHEDULE YOUR STUDY TIME

Writing assignments on your To Do List makes them seem more urgent, but that does not cover the daily work that must be done. You must take charge of your time and studying. In addition to specific assignments, each day you should:

- go over that day's notes,

- read the corresponding sections in the textbook,

- quiz yourself,

- review your notes again, and

- preview the next day's material.

All of this takes time. You must build study time into your schedule or you either will not get around to it or you will put it off until you are too tired to study effectively. The first thing to do is to write your study time into your day planner and master calendar, and regard that time as sacred—don't borrow from it to do something else. Be sure to allow break time during study sessions as well—if you study too long, your brain gets weary and starts to wander, and it takes much longer to do even simple tasks. Plan a 10- to 15-minute break for every hour of studying.

CHUNK IT

If a job seems too large, we put it off, but if we have many small tasks, each alone seems manageable. Break your workload into small chunks. Write them down, partly so you do not forget any, but especially because you will get a great feeling of accomplishment when you complete a task and cross it off your To Do List! Completing a task is also a convenient time to take a mini-break to keep your mind fresh. Many students try to read a whole chapter or cover a few weeks of notes in one sitting. The brain really dislikes that. When studying a large amount of material, divide it into subcategories, then study one until you really understand it before moving to the next.

STUDY ACTIVELY

Merely reading your notes or the book is not learning. You must think about the material and become an **active learner**. Constantly ask yourself, "What is most important in this section?" While reading, take notes or underline key terms and major concepts. Make flash cards. Consider how what you are studying relates to something with which you are already familiar. If you can put the information in a familiar context, you will retain it better.

The best preparation for quizzes and tests is practice. Develop and answer questions as you read. Try to anticipate all the ways your instructor might quiz you about that material. Recall which specific items your instructor stressed. Outline the material in each section and be sure to understand how the different concepts are related. Check yourself on the meanings and usage of the key terms. Say the key words out loud and look carefully at them. Do they remind you of anything? Have you heard them before? Can you spell them?

MOVE PAST MEMORIZING

This is one of the hardest study traps to avoid. In anatomy and physiology, it may at times seem like there is so much to learn and so little time. Most students at first attempt to just memorize. If you only read your notes and the book, you are using this approach without realizing it.

At the beginning of this section, I gave you three items to learn. Without turning back, write the names I asked you to learn a few pages ago:

1. _____

2. _____

3. _____

Did you remember them? Now, also without looking back, can you explain each of them to me? _____

_____

_____

_____

(I am betting not)

These three "things" are fictitious, but my point is that you may, indeed, have memorized the names—it doesn't take much to memorize words—but it takes a lot more to understand them, especially if the words are unfamiliar, as they often will be in this course. If you find that you study hard but the wording of the quiz or test confuses you, I can almost guarantee that you are memorizing. The question is worded a bit differently than what you memorized, so you don't realize that you know the answer. You must get past memorizing by looking for relationships between the concepts and terms, and really strive for full understanding. Reading often produces memorization. Active studying produces understanding.

THE CONCEPT MAP

A very useful technique for learning relationships is drawing a **concept map**. This is somewhat like brainstorming. Here is the general process:

1. Start with a blank piece of (preferably) unlined paper.

2. Near the center, draw a circle and, inside it, list the main concept you will explore.

3. Around that circle, and allowing some space, draw more circles and list in each anything that pops into your mind related to your main concept. Do this quickly and don't think about the relationships yet. Just get your ideas down.

4. Once you've added all your secondary concepts, look at them and think about how they are related, not just to the main concept but to each other as well.

5. As relationships occur to you, draw arrows connecting related concepts and add a brief description of the relationships between them.

6. Examine the relationships and you will start to understand how these concepts fit together.

You can also use concept maps to learn terminology. On blank paper, randomly write new terms from lecture or from a section of your textbook. Then think about the terms. What does each term mean? Which are related to each other? How are they related? Next, draw lines to connect related terms, adding brief explanations of the relationships between them. Try this with flash cards, with the term on one side and its definition on the other. Scatter the cards in front of you. Look at each term and quiz yourself on its meaning, then flip the card to check the definition. Flip the cards back over so the terms are facing up, then think about how the terms are related. As relationships emerge, move the cards around and place other cards between them on which you note the relationships. You might start by placing the cards with the definitions visible—sometimes it's easier to see relationships by comparing the meanings. Once you complete this step, turn the cards so you see the terms that are related, and then restudy their definitions and relationships.

TIME TO TRY

Construct a concept map around the main concept of *energy* by adding arrows to show relationships between the following concepts: cell activity, food, plants, the sun, and work.

When you are finished, look at **Figure 1.3.**

First draw a circle or "node" for each concept, keeping the main concept, if there is one, near the middle.

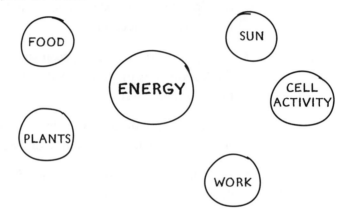

Next, add arrows linking the different concepts to each other, then add brief descriptions of how they are linked.

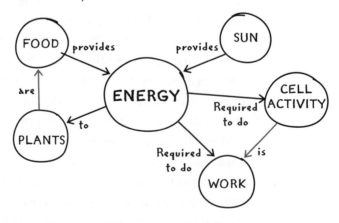

FIGURE 1.3   **Drawing a concept map.** In this example, the arrows show that the sun provides energy to the plants, and plants are food that provide energy required to do cell activity, which is a type of work.

REVIEW

OK, you've been at this study thing for awhile and you think you're starting to get the material. You did all of the tasks above. Can you quit now? Almost, but once you think you have the material under control, review it one more time. Repetition is the key to your long-term memory—the more you go over the material, the longer it will stay with you. There is actually a physiological basis for this—you are stimulating and reinforcing neural pathways in your brain. I always recommend a minimum of three passes, even for easy stuff—read your notes, read the text, re-read your notes—and that assumes you are understanding the material. Always slow down and go through it more if you are struggling with a certain section. Use **active learning** with each pass, then finish with one more review. If you are alert enough to review right before you go to bed, once you are asleep your brain often continues going over the material without bothering you too much (although one of my students reported a dream in which she was chased by a herd of giant bones!).

The website for your textbook also provides a good way to review. The address is in your book. Most book websites offer a wide variety of activity options, perhaps including online animations, flashcards, puzzles, objectives, vocabulary lists, and quizzes. When taking online quizzes, be sure to do so without looking in your book or your notes—after all, you do not use them for quizzes in class. Not using them provides a better simulation of the classroom experience. And, if you do well on the online quizzes without using your notes, you will have confidence in the classroom knowing you have already passed one quiz!

NO CRAMMING ALLOWED!

I have a very busy life, so my house occasionally gets a bit cluttered. If company drops by and it is a "bad house day," I might quickly grab some of the clutter and cram it into a spare closet. After the guests leave, perhaps I open the closet door to pull out a quilt. What happens? _____

_____

Now imagine what you do to your brain when you cram for an exam. You are essentially opening the closet door and cramming stuff in, then slamming the door. When you are taking the test, you open the door to pull out the answer you need, but anything might tumble out and land on your paper. Cramming at best allows partial memorization. At worst, it causes the information to get mixed up and you fail. It is a desperate act of superficial studying guaranteed to NOT get you through anatomy and physiology. If you study on a daily basis instead of doing a panicky cram session, before a test you will be calmly reviewing what you already learned well and smiling at the crammers in class.

NO VAMPIRES ALLOWED!

*Do you think you can pull an all-nighter and really do well?*_____
What do you think are some of the reasons this will not work? _____

_____

If you normally live your life by day, you can't suddenly override your natural biological clock and expect your brain to stay alert and focused late at night when it knows it should be asleep. Sleep deprivation impairs focus and attention. Your eyes may skim the page but you'll struggle to comprehend the words and retain very little of the small amount you do manage to absorb. Caffeine may temporarily keep your eyes open, but you'll only be a tad more alert while you still mentally drift away from the task at hand. Caffeine may also prevent sleeping later on.

Much of what we try to learn is solidified in our brains (moved to long-term memory) while we sleep, so without adequate sleep you won't retain what you tried to learn. And the next day, you won't be mentally alert while taking your test; you won't be able to focus or think things through. Your exhausted neurons won't be able to coordinate the information you tried to learn, you'll be less able to recall what you covered, it will be hard to understand exam questions, and your judgment will be impaired—doesn't sound like a good situation for testing, does it? Even sleep deprivation of as little as one hour can impair mental function the next day, so forget the all-nighter. Give your brain what it wants and needs—a good night of sleep.

An all-nighter is basically a marathon cram session held at the worst possible time. It simultaneously robs your brain and body of what they need—restoration before the next day. You may be able to stay awake all night, but if you doze off you may oversleep and miss your exam. Or, if you do arrive (I hope you weren't driving with no sleep!), you'll likely get part way into the test only to have your brain bail on you. If you are prone to "test anxiety," your defenses will be down and you will quite likely freeze and fail. Ah, if only you had been studying all along . . .

For your brain to be kind to you, you have to be kind to its home. You must take care of yourself physically—eat, sleep, exercise, and RELAX. ▪

✔ **QUICK CHECK**

Why should you study every day if the test is not for two weeks?

_____

_____

Strength in Numbers: **The Study Group**

One of the best ways to learn anything is to teach it to someone else, so form a study group. Although this may not be the best option for everyone, it is highly effective for many students. In fact, many instructors view study groups as essential for success in A&P. As soon as possible, start asking your classmates who wants to be in a study group—you *will* get people to join. You can quiz each other, discuss the material, help each other, and quite importantly, support each other. If you study solo, you may not be aware of your weaknesses. Your study partners can help you identify them and overcome them. A good way to work in a study group is to split up the material and assign different sections to different members, who then master the material and teach it to the group. Each member should also be studying it all on his or her own—that ensures better effort from everyone and allows other members to correct any errors in a presentation.

Answer: Studying on a regular basis breaks the material into smaller, more manageable pieces that you can master; the material is fresh in your mind and you will only need to review it before the test.

Scheduling joint study sessions can be challenging. Many students find that scheduling group sessions before or after class works best. You may want to establish some ground rules, including agreeing to use the time for studying and not for gossiping or just socializing. And although it may be tempting to meet over a pizza, you do want a quiet location where you can freely discuss the material with few distractions. Check with your instructor to see if there are any open lab times that might work for this. Some areas on campus may be available for study groups.

SQRHuh? **How to Read a Textbook**

Science textbooks do not read like novels, so you need to approach them differently. The name may sound odd, but **SQR3** is an effective method for studying your textbook. This method also works well for reviewing your notes. It stands for *S*urvey, *Q*uestion, *R*ead, *R*ecite, *R*eview.

During the **survey phase,** you basically skim the chapter. Read the chapter title, the chapter introduction, any other items at the beginning of the chapter, and all of the headings. This gives you the road map of where you will be going in the chapter. As you skim the chapter, also read all items in bold or italic. Next, read the summary at the end of the chapter.

During the **question phase**, look at the heading of each section and form as many questions as you can that you think may be covered in that section. Write them down. Try to be comprehensive in this step. By having these questions in mind, you will automatically search for answers as you read. You may also come up with some of the questions on which you will later be tested.

Now **read** the chapter for details. Take your time. Adjust your reading speed with the difficulty of the material. Also, keep in mind the questions you developed and try to answer them.

The next phase is to **recite**. You are now working on your ability to recall information. After reading each section, think about your questions and try to answer them from recall. If you cannot, reread the section and try again. Continue this cycle until you can recite the answers.

Finally, you want to **review**. This helps reinforce your memory. After you complete the previous steps for all sections you're studying, go back to each heading and see if you can still answer all of your questions. Repeat the recite phase until you can. When you are done, be sure you can also answer the questions at the end of the chapter.

SQR3 is one way to read a textbook, but there are many others. A similar method is called **PORPE.** With this method, after reading a section you:

- *Predict* possible essay questions.

- *Organize,* summarize, and synthesize the major points in your own words.

- *Rehearse* by reciting the information and quizzing yourself.

- *Practice* your answers to the essay questions you identified.

- *Evaluate* your work for accuracy and completeness.

Another method suggests a three-column note-taking system:

What I know	What I want to know	What I learned

In the first column, list what you already know before you read the topic. In the second column, write questions about the content that you want to answer. To do this, preview the section by reading titles and headings and examining tables and figures. Finally, read the material thoroughly and use the third column to answer your questions from column 2.

Each method is effective, but the best method for you is what *works* for you. Try each method and modify them to suit your needs. Incorporate your learning style preferences. Visit *Get Ready's* companion website at www.aw-bc.com/getready for some specific activities to try. The more of them you incorporate, the better you'll reinforce what you read. This approach may feel strange at first, but you'll quickly discover how effective active reading is for truly learning the material.

As mentioned earlier, a science textbook isn't written like a novel, so you won't learn by reading it like one—straight through once. Also, the writing style is different. Scientific writing can be rather tedious and challenging. Break the reading into sections, and do at least three passes for each: prereading, in-depth reading, and final review. Here are some final tips:

- Read within 24 hours of the lecture while the lecture is fresh in your mind.

- Read slowly—comprehension and retention matter, not speed. If you don't understand something, take a deep breath, slow down, and reread it until you do.

- Don't skip unfamiliar words—look them up and jot them down. You must know the language to understand the concepts.

- If you're stuck or your mind is wandering, do a quick review of what you've covered and then take a break. This lets you process what you've learned and allows you to return later, ready again to focus on new material.

- Don't underestimate the importance of reading assignments— they are a major component of your course and, with an effective reading method, they can be the key to mastering course content and succeeding in your course.

✔ **QUICK CHECK**

What is the minimum number of passes you should make for each section you read?

_____

_____

Answer: You should make at least three passes: preread (preview), read, and review.

A Place to Call My Own: **The Study Environment**

Briefly describe the location where you plan to do most of your studying.

_____

_____

_____

Now you know how to study effectively, but we often overlook WHERE to study. Your options may be limited, so you need to make the best of what you have. Ideally your study spot is somewhat isolated and free of distractions like TV, music, and people. At the least, you should minimize distractions.

Do you study in front of a TV that is on? Even if you try hard to ignore it, you will be drawn to it, especially if the material you are studying is tough. Music can be tricky—songs that you know, especially peppy ones, may get you tapping and singing along with them while you think your mind is actively engaged in learning. However, soft or classical music can keep you calm and more focused, unless you really dislike that kind of music.

Thinking about the study site you listed above, what distractions might you face? _____

_____

_____

_____

How can you minimize them? _____

_____

_____

_____

_____

If you cannot, what might make a better study spot? _____

_____

_____

For studying, you really need a space that is your own. A desk is a good place (unless it is also the computer desk shared by other members of your household or at which you spend hours playing computer games for fun!). Ideally it will be a place where you do nothing but study, so that when you are seated there you know exactly what your purpose is. If you are having trouble staying on task in your work spot, get up and walk away briefly. The mental and physical break may help you "come back" to work, and you won't begin associating the spot with struggling. Your study spot should be quiet and it should have good lighting to avoid eye strain, a comfortable chair, good ventilation and temperature, and a work surface on which you can spread out.

You will spend a lot of time studying here, so take time initially to set up your study space. The area should be uncluttered and well organized. It should also be inspiring and motivational. Perhaps frame a sign that says "*I WILL be a _____(your career)_____ by __(your goal date)__*." Try this same approach for the goals you set at the beginning of this chapter, and display them boldly and prominently. Consider displaying a photo of your hero, or of someone in the family who inspires you or whom you want to make proud. Parents, you might display a photo of your children with a caption saying something like "You are my reason," or "I will teach by example." Realize that you are their role model—your children's attitudes toward education will be formed by what you do now. This also applies for those of you who have brothers or sisters. You may be *their* inspiration. With these treasures surrounding you, you're a mere glance away from being reinvigorated if a study session starts to fizzle.

WHY SHOULD I CARE?

Most of your learning is done outside of class. The more efficiently you study, the better you will learn. Your study spot affects your attitude and concentration. The more seriously you take your study location, the more seriously you will study there.

If you live with family or a roommate, you absolutely must stress to them the importance of respecting your study time and study space. Be sure they know your career goals and why they are important to you, and ask them to help by giving you the time and space you need to succeed. Ask them not to disturb you when you are in your space. If you have small children who want time with mommy or daddy while you are studying, assuming they have adequate supervision, try getting them to play or study on their own until "the clock hands are in these positions," then do something fun with them at that time. They will learn to antici- pate your time together and to leave you alone if the reward is worth- while. This will also provide you with a relaxing study break. If you have too many distractions at home, the solution is to study elsewhere. Whether on campus, in the local library, or at a friend's house, you need a distraction-free setting, and if you can't get it at home, remove yourself instead of trying to cope with a poor study space.

✔ **QUICK CHECK**

What are some of the main considerations in selecting your
study spot? _____

_____

My, How **Time** Flies!

You know you need to study and that it takes a lot of time, but how will you fit it in? Let's discuss a few ways to budget time for studying. First, be consistent. Consider your schedule to see if you can study at the same time each day. Studying will become a habit more easily if you always do it at the same time. Some students adhere to one schedule on weekdays and a different one on weekends. When scheduling study time, consider your other obligations and how distracted you might be by other peo- ple's activities at those times. Don't overlook free hours you might have while on campus and even consider building some into your schedule. Head to the library, study room, or a quiet corner. This is the ideal time to preview for the next class or to review what was just covered.

Answer: Few distractions, own space just for studying, comfortable, good light-
ing/ventilation, sufficient work space, and welcoming.

TIME TO TRY

This is a two-part exercise designed to help you find your study time.

Part A: Each week has a total of 168 hours. How do you spend *yours?* **Table 1.4** on the facing page allows you to quickly approximate how you spend your time each week.

1. Complete the assessment in Table 1.4 to see how many hours are left each week for you to study.

2. Enter that number here. _____ hours

Part B: Next, turn your attention to **Table 1.5** on page 38.

1. Enter your class schedule, work schedule, and any other activities in which you regularly participate.

2. Now look for times when you can schedule study sessions and write them in.

3. Are you able to schedule 2 to 3 hours of study time per hour of class time? _____

It can be difficult, but it is essential to make the time. Writing it into your schedule makes it more likely to happen.

Don't overbook! Be sure to build in break time during and between your study sessions, especially the longer ones. Allow for flexibility—realize that unexpected events occur, so be sure you have some extra time available. Also, be sure you plan for and schedule recreation, too. You cannot and should not study all the time, but these other activities do take time and need to be in your schedule as well so that you do not double-book yourself.

TABLE 1.4   **Assessing how your time is spent.** For each item in this inventory, really think before answering and be as honest as possible. Items that are done each day must be multiplied by 7 to get your weekly total. One item may be done any number of times a week, so you'll need to multiply that item by the number of times each week you do it. After you have responded to all the questions, you'll have an opportunity to see how many hours remain during the week for studying.

Where Does Your Time Go? Record the number of hours you spend:	How many hours per day?	How many days per week?	Total hours per week: (hours × days)
1. **Grooming,** including showering, shaving, dressing, makeup, and so on.			
2. **Dining,** including preparing food, eating, and cleaning up.			
3. **Commuting** to and from class and work, from door to door.			
4. **Working** at your place of employment.			
5. **Attending class.**			
6. **Doing chores** at home, including housework, mowing, laundry, and so on.			
7. **Caring** for family, a loved one, or a pet.			
8. On **extracurricular activities** such as clubs, church, volunteering.			
9. **Doing errands.**			
10. On **solo recreation,** including TV, reading, games, working out, and so on.			
11. **Socializing,** including parties, phone calls, hanging out, dating, and so on.			
12. **Sleeping** (don't forget those naps!)			
Now add all numbers in the far column to get the total time you spend on all these activities.			
		Hours/week	168
		Total hours spent on other activities –	
		Left for studying =	

TABLE 1.5 My study schedule.

Time	Monday	Tuesday	Wednesday	Thursday	Friday	Saturday
6:00 AM						
7:00 AM						
8:00 AM						
9:00 AM						
10:00 AM						
11:00 AM						
Noon						
1:00 PM						
2:00 PM						
3:00 PM						
4:00 PM						
5:00 PM						
6:00 PM						
7:00 PM						
8:00 PM						
9:00 PM						
10:00 PM						
11:00 PM						
Midnight						

Putting It to **The Test**

If I had a nickel for every student who said they have test anxiety . . .

Some people really do suffer from true test anxiety, but the majority of students who claim to have this condition believe it to be true not because of an actual diagnosis, but rather because they get very nervous and may go blank during tests. If I ask a class who amongst them suffers from test anxiety, most hands go up. By the end of the semester, with some coaching, the number is far less. Why? They have learned how to take tests and how to stay calm. If you do suffer from true test anxiety, consult with your counselor right away so he or she can put you in contact with the support services you need to understand your condition and to learn how to conquer it. **Table 1.6** offers some tips on ways to reduce your anxiety about tests.

Most people dread taking tests and experience some anxiety when taking them. Not surprisingly, the better prepared you are for an exam, the less worried you will be. The best remedy for the stress you associate with taking tests is to be very well prepared. If you know you understand the material, what is left to worry about?

Some people get very anxious before tests because they fear they will not do well. This may be because they know they are not prepared. Again, the remedy is simple: Study well. But anxiety can also arise from a bad past experience. If you have done poorly on tests in the past, your self-confidence may be shot, so you anticipate doing poorly. That may lead to cramming and memorizing instead of truly learning, and may cause you to become excessively nervous during the test, which can cause poor performance. All you need is a couple of good grades on tests to get your confidence back! And effective studying will help you get those grades.

If you are a nervous test taker, stop studying for at least an hour immediately before your exam. Most of my students who complain of test anxiety are frantically reviewing their notes right up to the moment I give them the test. They have been trying to quickly glance back over everything while racing against the clock. No wonder they are stressed when they begin the test! Remember that your brain needs time to process the information. When you cram information into the "closet," who knows what will fall out when you open the door during the test.

TABLE 1.6   **Ways to minimize your test stress.**

When	Actions
While preparing for the test	<ul><li>Study daily to avoid last-minute cramming.</li><li>Start reviewing several days before big tests.</li><li>Quiz yourself on terms.</li><li>Review concept maps.</li><li>Review the questions you developed while reading the material.</li><li>Read your notes for anything the instructor emphasized.</li><li>Consider possible essay questions and write out thorough answers.</li><li>Review materials and practice quizzes on your textbook's website.</li><li>Meet more often with your study group; focus on reviewing and quizzing.</li><li>Counter negative thoughts ("There's too much!") with positives ("I've studied hard and I know this stuff!").</li><li>Remind yourself this is just one grade—it won't determine your worth or your future.</li><li>Take adequate study breaks.</li><li>Eat well and exercise (a great stress reliever).</li><li>Avoid sleep deprivation, especially the night before the exam.</li></ul>
On the day of the test	<ul><li>Eat moderately; don't skip meals; don't eat anything too heavy to make you groggy.</li><li>Avoid excessive caffeine—it increases anxiety and makes you jittery.</li><li>Stop studying at least an hour before exam time and do something to relax.</li><li>Stretch out somewhere comfortable. Focus on fully relaxing your muscles—from your head to your toes—savoring the feeling of relaxation.</li><li>Listen to calming music before arriving in class.</li><li>Arrive early to get your seat and organize your thoughts (do NOT look at your notes!); put your head down and relax until test time.</li><li>Avoid discussing course content or listening to classmates before the exam.</li></ul>
During the test	<ul><li>Take five slow, deep breaths just before you begin.</li><li>Browse the whole test so you know its layout and length.</li><li>Budget your time, allowing more time for harder sections.</li><li>Carefully read all directions twice.</li><li>Start with the easiest part of the test.</li><li>Check the clock regularly to assess your progress and adjust your speed as needed.</li></ul> ▶

TABLE 1.6   Ways to minimize your test stress, continued.

When	Actions
During the test (continued)	■ If you start to feel anxious, close your eyes, focus on relaxing as you take slow, deep breaths, and remind yourself how well you prepared. ■ Focus on one question at a time—read it carefully, underline key words, and think before answering. ■ Ignore students who finish before you—often the first people to leave had little to write. Take your time and be thorough and careful. ■ If you run short on time, answer what you can quickly and just leave the rest.
After the test	■ Reward yourself regardless of how you feel you did. ■ Don't obsess over how you think you did—you'll know when the test is returned, and your "second" guesses are likely not accurate. ■ Review any material that you still feel unsure about. ■ Once returned, record your score and how many points the test was worth. ■ Examine your test. Note what you missed and why, and ask for clarification if you're not sure. Review that material again. ■ Let it go—it's finished and you can't change it now. Move on.

If you have studied well in advance and don't get very nervous at exam time, you might want to glance quickly through your notes beforehand, but only if you have time to do so and still allow *at least* a half-hour to relax and mentally prepare for your test. The half-hour off allows your brain to process the information while you relax. Try getting a light snack so you are alert—a heavy meal could make you drowsy during the test. Walk around to release nervous energy. Listen to music that makes you happy and relaxed. Sit comfortably, close your eyes, and breathe deeply and slowly while you picture yourself in a very relaxing setting—maybe on a tropical beach, curled up on your couch with a good book, or out on a boat fishing. Focus on how relaxed you feel and try to hold that feeling. Now, staying in that mood, concentrate on how well you have studied and keep reminding yourself that:

■ I have prepared very well for this test.

■ I know this material very well and I answered all questions correctly while studying.

▶

- I can and *will* do well on this test.

- I refuse to get nervous over one silly test and one grade, especially because I know I am ready.

- I am ready and relaxed. Let's get it done!

TEST-TAKING TIPS

There are also strategies you can use while taking the test. Let's explore your current strategies. Complete the survey in **Table 1.7**, then we will discuss specific strategies.

During an exam, be careful—read each question thoroughly before you answer. This is especially true of multiple choice and true/false questions. We know the answer is there, so our eyes tend to get ahead of our brains. We skim the question and jump down to the answers before even trying to mentally answer the question. Slow down and think before moving to the answers. Otherwise you may grab an answer that sounds familiar but is incorrect. An easy way to slow down and focus on the question is to underline key words as you read it. If you have trouble keeping your eyes off the answers, cover them with your hand until you finish reading the question, think of the answer on your own, then read all the answers to see if yours is there.

If you do not know the answer initially, take a deep breath and think of all you do know about the words in the question. Often this is all you need to recall the answer. This is when those concept maps you made will really come through for you.

Use the process of elimination. If you are not sure which answer is correct, can you eliminate any you know are incorrect? Narrow down your choices. Avoid making a guess unless the process of elimination fails you; however, guessing is usually better than leaving a question unanswered. On short-answer questions, fill-in-the-blank questions, and essays, always write something. Whatever you write just may be correct, but an empty space is always wrong.

After you answer a question, read your answer to be sure it says what you want it to, then leave it alone. Once you move on, avoid the temptation to go back and change your answers, even those of which you were unsure. Often we have a gut instinct to write the correct answer; perhaps we are recalling it at some subconscious level. But the very act

TABLE 1.7 **Self-evaluation of test-taking skills.** Think about how you have prepared for tests in your previous courses. For each of the following valuable test-taking skills, mark if you do each one always, sometimes, or never. Highlight any that you do not currently use that you think might help you be more successful.

Test-Taking skill	Always	Sometimes	Never
1. While studying my notes and the book, I think of and answer possible test questions.			
2. I use online practice quizzes when they are available.			
3. I avoid last-minute cramming to avoid confusing myself.			
4. I scan the whole test before starting to see how long it is and what type of questions it contains.			
5. I do the questions I am sure of first.			
6. I budget my time during a test so I can complete it.			
7. I answer questions with the highest point values first.			
8. I read all answer options on multiple choice questions before marking my answer.			
9. I know what key words to look for in a multiple choice question.			
10. I use the process of elimination during multiple choice or matching tests.			
11. I know what key words to look for in essay questions.			
12. I look for key words like *always, never,* and *sometimes.*			
13. When I am unsure of an answer, I go with my first answer and fight the urge to change it later.			
14. I try to answer everything even if I am uncertain, instead of leaving some questions blank.			
15. I check my answers before turning in a test.			

of going back is a conscious reminder of uncertainty, and we often choose something different only because we doubt ourselves.

When answering multiple choice or true/false questions, ignore any advice that suggests you should select one answer consistently over others. Also, don't worry if you choose the same answer several times in a row, thinking the instructor would not structure a test that way. I can't speak for all instructors, but I do not personally know any who give much thought to the pattern the answers will make on the answer sheet, so neither should you.

Here are a few more pointers:

- Glance over the exam as soon as you receive it, so you know what to expect, then budget your time accordingly.

- Look for questions on the backs of pages so you don't miss them.

- Note the wording on questions. Key words to look for that can change an answer are *always, sometimes, never, most, some, all, none, is,* and *is not.*

- Tackle easy questions first. They may provide hints to the tougher ones.

- Be aware of point values and be sure the questions with the greatest point values are done well. Often essay questions—which usually are worth more points—are at the end, and some students run out of time before reaching them, losing significant points and seriously hurting their grades.

- If you have trouble writing essay answers, recall all you know about the topic, organize in your mind how you would explain it to someone, then write down your thoughts as if you are writing yourself a script on what to say.

- If a question has multiple parts, be sure to answer each part. This is especially true for essays.

- For true/false questions, mark true only if the *entire* statement is true. If any part is false, mark it as false.

- For multiple choice questions, read the directions carefully—you may have to choose more than one answer.

- If you are asked to choose only one answer, choose the *best* answer. Although more than one answer may be correct, the most inclusive is the best answer.

- Look for answers elsewhere in the test.

- If the instructions say "... *include in your answer...*," then DO!

- If you are asked for a definition, give a book explanation of what the term or concept means. If you are asked for an example, list an example and explain why it is an example of the concept. If you are asked to explain a concept or term, approach it as if you are trying to teach it to a 12-year-old. Assume the grader has no prior knowledge.

- Be very thorough and specific in your answers. The grader cannot get inside your head to decide if you knew it or not, so your words must very literally convey your meaning.

When a test is returned, record your grade. Be sure to review the test to see which questions you missed and why you missed them, and make notes to go back and review that material. Remember—it may come back to haunt you on a bigger test or on the final exam, and you should know it anyway.

✔ **QUICK CHECK**

How can you slow yourself down when taking a multiple choice or true/false test? _____

_____

Through the Looking Glass: **Individual Accountability**

I hope you have gained insight into the learning process and developed new strategies to improve your success, not just in anatomy and physiology, but in all of your classes. One area remains for us to discuss, though, and that is your responsibility and attitude. When we get frustrated, we often look elsewhere for the cause, even when it may be right on top of our own shoulders. I have watched poorly prepared students transform into honors students and I have seen honors students drop out as they start getting really bad grades. Many factors can contribute to these changes, but a common thread is always attitude and accountability. Here are three facts you need to firmly implant in your mind:

1. *You*, and nobody else, chose to pursue this academic and career path.

2. *You*, and nobody else, are responsible for attaining the success you desire.

3. *You*, and nobody else, earn the grades you get.

Answer: Cover the answers with your hand while you read the question, and don't look at them until you think of the answer.

You must do everything you can to guarantee your success—nobody will do it for you. That means always accepting responsibility for your own actions and effort. No excuses. To stay on track, you must know exactly where you are at all times. You must always know precisely what your grade is (we will review how to calculate your grade in the math chapter) and remember that you always get the grade that *you earn* through your hard work (or lack of it). You must always be very clear about exactly what you want and always stay focused on where you are going. At times, you may not feel like you can keep up, but instead of quitting or slacking off, you need to refocus on where you are going and why it matters to you. Always set short-term and long-term goals. Write them down and post them where you will see them often. You are responsible for keeping yourself motivated. Learn to visualize your success—see yourself in your future career. Think about how your life will be and how that will benefit those around you. Dream big. Then go after that dream with all you have.

An important part of any journey is to anticipate roadblocks before you hit them. Carefully think of any possible obstacles to your success, then plan around them. You have an unreliable car? Find someone you can ride with in emergencies or look into public transportation. You have small children at home? Have daycare lined up and a backup plan for when your child is too ill to go to daycare. You have a learning disability? Immediately contact student support services or your counselor to find out what services are available to assist you. Make a list of anything that might get between you and your success, then write down at least two solutions so you have a main strategy and a backup plan.

Finally, consider those around you. Family and friends must know your goals and understand how important they are. But do they support you? I have had women whose husbands burned their books because the men felt threatened that their wives might no longer need them once a degree was attained. And there are certainly more subtle means of sabotage. Perhaps your friends needle you because you don't go out as much, or they say you're no fun anymore. Your significant other whines that you don't get enough time together. Relatives accuse you of thinking you are better than them because you are getting some education.

Realize that when you change, whether through education or something else, those who know you may feel excluded, threatened, left behind, or even envious. You can try to assure these people how much you still value and need them in your life, but don't let them distract you from your mission.

You must surround yourself with supportive people who are happy and proud of you, who celebrate your victories, and who want for you what you want. They will help you succeed. It may be your study group or others around you who will help you study. Perhaps they will watch your children so you get some quiet time. On the other hand, anyone who ridicules you or is upset by your new schedule and goals is really not a friend you want around. Make new friends in class who share your goals and guard yourself from those who would derail you. Stay on track, and, if you *choose* to, you will soon be living your dream. Good luck!

Final Stretch!

Now that you have finished reading this chapter, it's time to stretch your brain a bit and check how much you learned. For online tests, tutorials, animations, activities, web links, and an ebook, visit the *Get Ready for A&P* companion website.

RUNNING WORDS

At the end of each chapter, be sure you have learned the language. Here are the terms introduced in this chapter with which you should be familiar. Write them in a notebook or enter them in your computer. Define them in your own words, then go back through the chapter to check your meaning, correcting as needed. Also try to list examples when appropriate.

Visual learner
Auditory learner
Tactile learner
SMART
Active learner
Concept map
Active learning
SQR3
PORPE

WHAT DID YOU LEARN . . .

Part A: In the left-hand column below, write your approach before reading this chapter. In the right-hand column, list any changes you plan to make to ensure your success in this class.

What I have done before this chapter	What I will do to improve
During lectures:	
Note-taking:	
Study habits:	
Textbook reading:	
My study place:	
Time management:	
Test-taking:	

Part B: List the three areas in which you think your study skills are the weakest, and ways in which you plan to improve them.

1.

2.

3.

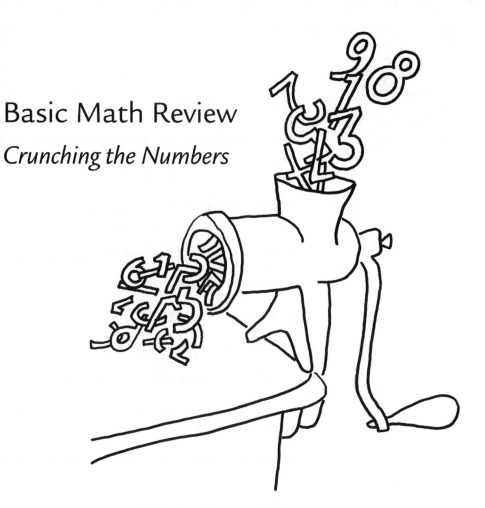

2 Basic Math Review

Crunching the Numbers

When you complete this chapter, you should be able to:

■ Solve math problems involving multiple operations, fractions, decimals, and percents.

■ Calculate the mathematical mean and explain its importance to physiology.

■ Work with exponents, numbers in scientific notation, ratios, and proportions.

■ Understand the units of the metric system.

■ Perform basic measurements in a laboratory setting.

■ Read tables, graphs, and charts.

Your Starting Point

Answer the following questions to assess your math skills.

1. $2/3 \times 3/4 =$ _____

2. Express 50% as a fraction: _____
   As a decimal: _____

3. What is the *mean* of 27, 33, and 36? _____

4. In scientific notation, 1000 = 10 —

5. 30% of 200 = _____

6. Which is longer, 1 yard or 1 meter? _____

7. In the metric system, the basic unit of volume is the _____.

8. Assume you take 15 quizzes by the end of the semester and get 9 A's and 6 B's. Express this as a ratio. _____

9. Is a triple beam balance used to measure volume, mass, or distance?
   _____

10. On a graph, the vertical line is the _____ axis.

How Much Wood Would a Woodchuck Chuck? **Math in Science**

You probably remember doing story problems when learning math in your younger years. Those problems helped you see how math can be used. Many students are surprised to learn that they have to use math in anatomy and physiology, but you must remember that science—all science—deals with that which is testable. A scientific test, as you know, is called an experiment. Results collected from experiments are called **data** and, more often than not, the data are numbers. When you try to make sense of the data, you are working with numbers, and that means math.

You will likely do some experiments in class and then analyze the data. In addition, many aspects of body function have "normal" conditions that are often expressed in numerical values. For example, normal body temperature is 98.6°F, normal blood pressure is 120/80, and normal pulse is around 70 to 80 beats per minute. You will calculate various physiological values. For example, the amount of blood that leaves the heart each minute is called cardiac output, which is calculated by multiplying the number of heartbeats per minute (heart rate) by the amount of blood leaving the heart with each beat (stroke volume). You will work with chemical solutions and you will need to understand their concentrations. You will measure in metric units, refer to percentages and ratios, and interpret graphs and charts.

Many students entering this class need to dust off their math skills a bit. Some of you may only need a brief reminder of what you learned before, others may need to learn it again. Regardless of your math history, a quick refresher will help you better understand the numbers.

From the Beginning: **Basic Math Operations**

You might need to do some complicated computations in class, so it is good to remember the basic rules. Let's zip through them for a quick refresher. I assume you can add and subtract, so we'll skip those operations.

MULTIPLICATION   Multiplication problems can be done in any order: $3 \times 4 \times 2 = 24$, or $2 \times 3 \times 4 = 24$. The answer to the equation is called the **product**. Recall that multiplying any number by 1 does not change the number, while multiplying any number or numbers by 0 gives you 0.

In more complicated equations, an operation may be set off in parentheses or brackets. If a number appears immediately to the left of a parenthesis or bracket, multiply by that number even though there is no multiplication sign:

$$3(6-2) = 3 \times (6-2) = 3 \times 4 = 12$$

Multiplication problems are sometimes represented with **exponents**. Consider 2^4, which is read as "two to the fourth power," and 10^3, which is read as "ten to the third power." These examples are really just a short-hand way of expressing these multiplication problems:

$$2^4 = 2 \times 2 \times 2 \times 2 = 16 \qquad 10^3 = 10 \times 10 \times 10 = 1000$$

DIVISION Like subtraction, division problems must be done from left to right. Consider this example:

$$10 \div 5 = 2,$$
$$\text{but } 5 \div 10 = 1/2$$

In division, the number being divided is called the **dividend**, the number by which it is divided is the **divisor**, and the total is the **quotient**. In our first example, 10 is the dividend, 5 is the divisor, and 2 is the quotient.

MULTIPLE OPERATIONS   I am sure you remember all of this, but let's try some more complicated problems that involve more than one mathematical operation. Parentheses or brackets are often used in equations when there are multiple operations. Think of them as directors telling you how to proceed—always do operations within these structures first. Let's see why this matters.

$$8 - (2 \times 3) = 8 - (6) = 2,$$
$$\text{but } 8 - 2 \times 3, \text{ done in that order, } = 6 \times 3 = 18$$
$$\text{OOPS!}$$

Always approach an equation by first doing any operations within parentheses. ▪

Here are some simple rules to help ensure that you perform mathematical operations in the correct order.

1. First, do all operations inside the parentheses or brackets.

2. Next, multiply out any exponents.

3. Then do all multiplication and division equations, moving from left to right.

4. Finally, do all addition and subtraction problems, again from left to right.

TIME TO TRY

Are you with me so far? Try the following problems.

1. $4 \times (9 - 6) + 10 =$ _____

2. $3^3 \div 9 - 4 + 5 =$ _____

3. $6^2 - 2(5-2) + 4 - 2 =$ _____

Let's see how you did.

Problem #1: The correct answer is 22. First, do what is in the parentheses (rule 1): $(9 - 6) = 3$, so the problem becomes $4 \times (3) + 10$. Next, do the multiplication (rule 3): $4 \times 3 = 12$, so the problem becomes $12 + 10$. Finally, do the addition (rule 4): $12 + 10 = 22$.

Problem #2: The correct answer is 4. There are no parentheses, so you start with the exponent (rule 2): $3^3 = 3 \times 3 \times 3 = 27$, and the problem becomes $27 \div 9 - 4 + 5$. Next, do the division (rule 3): $27 \div 9 = 3$, so the problem becomes $3 - 4 + 5$. Finally, do the addition and subtraction from left to right (rule 4), and you get $3 - 4 = -1$, then $-1 + 5 = 4$.

Problem #3: The correct answer is 32. Start with the parentheses (rule 1): $(5 - 2) = 3$, so the problem becomes $6^2 - 2(3) + 4 - 2$. Next, take care of the exponent (rule 2): $6^2 = 6 \times 6 = 36$, so the problem becomes $36 - 2(3) + 4 - 2$. Now, do the multiplication and division from left to right (rule 3): $2(3) = 2 \times 3 = 6$, so the problem becomes $36 - 6 + 4 - 2$. Finally, do the addition and subtraction from left to right (rule 4): $36 - 6 = 30$, then $30 + 4 - 2 = 32$.

As you see, some mathematical equations can be long and somewhat complicated, but if you keep the basic rules in mind and tackle those problems step by step, they become quite manageable.

What Do You **Mean** You Are Normal?

Do you know what "normal" body temperature is? Sure you do—98.6°F. You have likely known that since you were a small child. But what is YOUR normal temperature? Mine is rarely 98.6°F, and yours may not be either. "Normal" blood pressure is 120/80, yet mine often runs lower than that. Just when I started thinking I might be abnormal, I reconsidered what the term "normal" really means.

In physiology, the term **normal** means **average**. And in math, another term for average is **mean**. We refer to many normal values—temperature, blood pressure, pulse, respiratory rate . . . the list goes on and on. When you see these, realize that they are average values and any individual may have a different normal value—what is normal for him or her may not be average for the whole population. To better understand this idea of normal, you need to know how to calculate the average, or mean, value.

Let's say you're doing a lab on the cardiovascular system and you're measuring pulse. You are instructed to do three trials and then calculate the mean pulse. Your three trials give you the following data:

> Trial 1: 72 beats per minute
> Trial 2: 74 beats per minute
> Trial 3: 79 beats per minute

To find the mean of a group of numbers, simply add them all together then divide the total by how many numbers you added. For your data, you would add the three pulse rates, then divide by 3:

$$72 + 74 + 79 = 225 \qquad 225 \div 3 = 75 \text{ beats per minute}$$

Did you notice that the mean is not one of the original numbers? It does not have to be. It is the average of all three numbers.

Here is a tip to help you with means and with most math problems—learn to predict your results! The mean of a group of numbers will be somewhere between the highest and the lowest of the numbers you are averaging. If your value does not fall in that range, check to see if you made an error. Common errors are missing a number during the addition or dividing by the wrong number. If you estimate your result first, you can more easily recognize errors if they occur.

TIME TO TRY

Calculate the mean of these body temperatures: 97.4°F, 98.0°F, 99.2°F, and 99.8°F.

1. What is the mean? _____ °F

2. What does this mean *mean*? _____

_____

If you did this correctly, you should have gotten a mean of 98.6°F, even though that was not one of the original temperatures listed. So, as stated earlier, the physiological "normal" value is a mean, and individuals' normal temperatures will vary around that mean.

WHY SHOULD I CARE?

All science is based on data and experimental trials. There is a certain amount of error possible with each trial. Consider the pulse values we used as examples. Three trials gave us three results. Using the mean helps to minimize the error from individual trials.

In physiology, means, or normal values, are important because they give us a reference point for how the body is working. If someone's temperature is 101.7°F, we are pretty sure there is a problem because it is significantly far from the normal value. We use means for comparison—the mean tells us what "normal" function should be so we can recognize abnormal function.

✔ QUICK CHECK

What is meant by saying that normal human heart rate is 80 beats per minute?

Meet My Better Half: **Fractions, Decimals, and Percents**

Working with whole numbers is rather easy and is second nature to most of us. Fractions, however, are often a faded, distant memory. Related to fractions are two other ways of expressing values: decimals and percents. In anatomy and physiology, all three of these will be used. For example, the micrometer (μm), a common unit for measuring the size of microscopic structures, is a tiny fraction of the more familiar millimeter—a micrometer is 1/1000 of a millimeter, to be exact. As you just saw, normal body temperature is reported as a decimal (98.6). Finally,

Answer: It means the average, or mean, heart rate for humans is 80 beats per minute.

about 60% of the body of an average adult male is water. You will discover that many physiological values are reported in any of these formats, so you want to be comfortable with their use.

FRACTIONS

Fractions are written as *a/b*, in which *a* and *b* are both whole numbers and *b* is not 0. The top number is called the **numerator**, and the one on the bottom is the **denominator**. You probably realize that a fraction gets larger as the numerator gets larger—it's obvious that 1/3 is smaller than 2/3. In contrast, as the denominator gets larger, the number gets smaller—for example, 1/2 is larger than 1/4, which, in turn, is larger than 1/8.

A fraction represents parts of some whole group (**Figure 2.1**). For example, 3/4 represents 3 equal parts out of 4 equal parts, where the 4 equal parts make up the whole (Figure 2.1a). Whole numbers can be represented as fractions as well. The whole number simply becomes the numerator, and the denominator is 1, so 3 = 3/1.

✔ **QUICK CHECK**

For 5/8, what is the numerator? _____ The denominator? _____
Express 6 as a fraction: _____

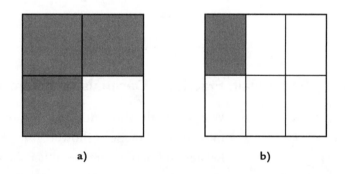

a)   b)

FIGURE 2.1 **Fractions represent some part of a whole.** In each of these squares, the shaded area represents the fraction that is listed below. **a)** This square is divided into four equal parts, and 3 of the four are shaded = 3/4. **b)** This square is divided into six equal parts, and 1 of the 6 is shaded = 1/6.

Answers: 5 is the numerator, 8 is the denominator, and 6 as a fraction is 6/1.

REDUCING FRACTIONS   **Equivalent fractions** have the same value even though they appear to be different. Consider the following fractions:

1/3   2/6   4/12   7/21

All of these numbers have the same value: 1/3. To see this, you need to **reduce** the other fractions. This is done by finding the **greatest common factor (GCF)** for each fraction. The greatest common factor is the largest whole number that can be divided into both the numerator and the denominator. Consider 2/6. Both the numerator (2) and the denominator (6) are divisible by 2, which is the greatest common factor. If you do the division, you see that $2 \div 2 = 1$ and $6 \div 2 = 3$, so 2/6 becomes 1/3.

TIME TO TRY

Look at the other fractions we listed: 4/12 and 7/21.

What is the greatest common factor for 4/12? _____

Divide the numerator by that factor: _____
Divide the denominator by that factor: _____

What is the reduced fraction? _____

What is the greatest common factor for 7/21? _____

Divide the numerator by that factor: _____
Divide the denominator by that factor: _____

What is the reduced fraction? _____

How did you do with Time to Try? You should have found that the greatest common factor for 4/12 is 4, so $4 \div 4 = 1$ and $12 \div 4 = 3$. Thus, the fraction 4/12 reduces to 1/3. Similarly, 7/21 has a greatest common factor of 7, and $7 \div 7 = 1$, and $21 \div 7 = 3$ so, again, 7/21 reduces to 1/3. After using a number to reduce the fraction, check your result to see if it is in its simplest form or if it can be reduced further.

✔ QUICK CHECK

What is the most reduced form of each of the following fractions: 60/90, 25/100, and 18/54?

MULTIPLYING AND DIVIDING FRACTIONS   When doing mathematical operations with fractions, the rules are different for multiplication and division than they are for addition and subtraction. For multiplication, you simply multiply the numerators in one step, then multiply the denominators. Consider 2/3 × 3/4. The numerators are 2 and 3. Multiply them to get 6, and that goes on top. Next, multiply the two denominators, 3 × 4, to get 12. So the product is 6/12, which reduces to 1/2:

$$\frac{2}{3} \times \frac{3}{4} = \frac{6}{12} = \frac{1}{2}$$

Let's try another: 1/3 × 2/5 × 3/4 = _____

First, multiply all the numerators (1 × 2 × 3 = 6), then multiply all the denominators (3 × 5 × 4 = 60) and you get 6/60, which reduces to 1/10.

To multiply fractions, first multiply all the numerators, then multiply all the denominators. Reduce the result as needed. ▪

Dividing fractions may seem difficult at first, but a simple trick actually makes it easy! These problems may be written two different ways:

$$\frac{4/5}{2/3} \text{ or } 4/5 \div 2/3$$

Solving them is easy. First, invert (flip) the second fraction, which is the divisor: 2/3 becomes 3/2. Then you simply multiply the two fractions:

$$4/5 \div 2/3 = 4/5 \times 3/2 = 12/10$$

Now, 12/10 can be reduced to 6/5. Here the numerator is larger than the denominator, which means this fraction is greater than 1. Usually when this happens, it is best to express the answer as a mixed number—one combining both whole numbers and fractions. To do this, first reduce the fraction: 12/10 = 6/5. Then realize that 6/5 = 5/5 + 1/5. Since 5/5 equals 1, the mixed number would be 1⅕.

To divide one fraction by another, first invert the second fraction to turn it into a multiplication problem. Next, multiply the numerators, then multiply the denominators. Finally, reduce the result. ■

ADDING AND SUBTRACTING FRACTIONS   To add or subtract fractions, they must first be in the same format. You might think you can just add the numerators and denominators, but that won't work. By that method, 1/2 + 1/4 would equal 2/6, which reduces to 1/3. But that is smaller than 1/2, one of the numbers we added! This doesn't make sense. (See why it helps to predict your results?) Instead, you must first put the fractions into common terms. They must have the same denominator, called a **common denominator**.

To get the common denominator, you need to know the **least common multiple (LCM)**. This is the smallest number that can be divided by both denominators. In our example of 1/2 + 1/4, 4 is the least common multiple, so we want both fractions to have 4 as their denominator. Recall that any number multiplied by 1 does not change. To convert 1/2 into fourths, we multiply it by 2/2 (=1). Thus, 1/2 x 2/2 becomes 2/4. Once the fractions have a common denominator, we simply add the numerators only:

$$2/4 + 1/4 = 3/4$$

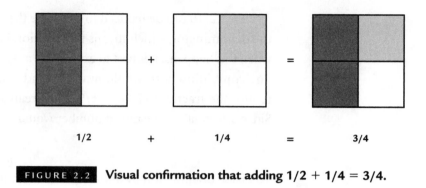

1/2   +   1/4   =   3/4

FIGURE 2.2   Visual confirmation that adding 1/2 + 1/4 = 3/4.

Figure 2.2 illustrates this for you.

Subtracting fractions also requires a common denominator. Once you have the denominator, simply subtract the second numerator from the first numerator. Let's try one: 1/3 − 1/4. The common denominator for these two fractions is 12, so they both need to be converted, as follows:

$$1/3 \times 4/4 = 4/12$$
$$1/4 \times 3/3 = 3/12$$

Now line up your fractions in the correct order from left to right, then subtract the second numerator from the first:

$$4/12 − 3/12 = 1/12$$

To add or subtract fractions, use a common denominator to put the fractions in a common form, then add or subtract the numerators only. Remember to always subtract from left to right. ■

✔ **QUICK CHECK**

Solve these problems:

1. 3/5 + 1/4 + 1/10 = _____

2. 9/16 − 3/8 = _____

Answers: 1. The common denominator is 20, so 3/5 × 4/4 = 12/20, 1/4 × 5/5 = 5/20, and 1/10 × 2/2 = 2/20. Then add the numerators: 12 + 5 + 2 = 19/20. **2.** The common denominator is 16, so 3/8 × 2/2 = 6/16. Then 9/16 − 6/16 = 3/16.

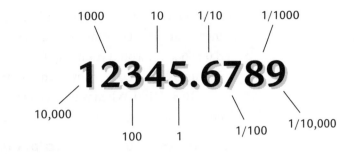

FIGURE 2.3 Each space around a decimal point reflects a change by a factor of 10.

DECIMALS

Decimals are common in science, and anatomy and physiology are no exception. All decimals are based on 10 in a very specific way—each place in the number represents a multiple of 10. The value *increases* 10 times for each space that you move to the left of the decimal, and it *decreases* 10 times for each space to the right.

Let's consider this number: **12345.6789**. You know how to read the numerals to the left of the decimal. They make 12,345. Moving from the decimal to the left, you can see how the spaces represent, in order, 1's, 10's, 100's, 1000's, and 10,000's (**Figure 2.3**). Our number represents 10,000 + 2000 + 300 + 40 + 5. The spaces to the right of the decimal represent fractions: tenths, hundredths, thousandths, and so on. Thus in our number, as you go to the right of the decimal, the numerals represent the fractions 6/10, 7/100, 8/1000, and 9/10,000.

CONVERTING DECIMALS   As with fractions, decimals allow you to express a number more precisely than you can with whole numbers. In fact, you can think of fractions as division problems that give you a decimal value, so fractions can be converted to decimals. For example, 1/2 is 1 ÷ 2, and that equals the decimal 0.5. Note the 0 here—it is used as a place holder so that you know where the decimal belongs. If you use a calculator to find 1 ÷ 3, the answer will be 0.333333 . . . . This is known as a **repeating decimal**. When you have a repeating decimal, one option is to round off. If the last number is less than 5, round down; if it is 5 or higher, round up. For example, 0.33333 . . . would round down to 0.33, and 0.66666 . . . would round up to 0.67.

Decimals can also be converted into fractions. The value 0.25 represents 2/10 + 5/100. When you add them, remember to first get a common denominator. You get 20/100 + 5/100 = 25/100, and that reduces to 1/4.

ADDING AND SUBTRACTING DECIMALS   Adding and subtracting decimals is easy as long as you line them up correctly. Consider this example: 1.287 + 24.32. First, be sure the two numbers have the same number of spaces after the decimal. The first number has three places, but the second has only two. But, you can always add zeros to the end of a decimal number, 24.32 = 24.320. Note that the last 0 means that there are 0/1000, which is correct and does not change the value. Finally, it is easiest to add these numbers if you line them up vertically, always being sure the decimal points are aligned:

$$
\begin{array}{r}
1.287 \\
+\ 24.320 \\
\hline
=\ 25.607
\end{array}
$$

Subtraction is done the same way.

PICTURE THIS

Assume that you have worked three extra jobs for pocket cash this week. From them, you earned $33.70, $45.28, and $21.02. How much extra money did you earn? _____

With this money you buy a pizza for $8.99, a soft drink for $1.49, gas for $20.00, and a new CD for $19.95. How much money do you have left? _____

Congratulations, you just added and subtracted decimals, as you do on a regular basis in daily life! You should see that you earned $100 and have $49.57 left.

When adding or subtracting decimals, always align the decimal point in the two numbers before doing the operation. ■

MULTIPLYING AND DIVIDING DECIMALS   Multiplication and division of decimals is a bit trickier because you must keep track of how many decimal places (digits to the right of the decimal point) you should have at the end. Let's try an easy one: 0.5×0.3. First, multiply the numbers as if they are whole numbers: $5 \times 3 = 15$. Now, add the number of decimal places you started with. Both numbers you multiplied originally had one decimal place, so that adds up to two. Realize that your answer of 15 is 15.0, so you know where the decimal begins. Now you have to move the decimal. The numbers you started with had a total of two digits after the decimal, so you must end up with two decimal places. You need to move the decimal point left by two places, giving you 0.15. Here is a way to double check that. If the original numbers were fractions, they would be 3/10 and 5/10. Recall how to multiply fractions—you multiply the numerators, then multiply the denominators:

$$3/10 \times 5/10 = 15/100 = 0.15$$

What if the numbers had been 0.03×0.5? Although you still get 15, now you need to move the decimal three places to the left, but there are only two. You simply add zeros to the left until you have the correct number of decimal places, in this case giving you 0.015.

Division with decimals is just like ordinary division, except we keep going until we either finish or reach a predetermined stopping point. You usually do not go beyond the number of decimal points your original numbers contain, so if they had a total of two, you would likely stop at two and round off beyond that. Let's look at an example: $2.8 \div 7$. Put the decimal point in the answer line exactly above its position in the dividend (2.8), then simply do the division:

$$
\begin{array}{r}
0.4 \\
7 \overline{)\ 2.8} \\
-\underline{2\ 8} \\
0
\end{array}
$$

How do you divide when both numbers are decimals? All you have to do is move the decimal of the divisor until you have a whole number, then move the decimal of the dividend by the same number of spaces in the same direction. If you try 1.68 (dividend) ÷ 0.3 (divisor), you move the decimal in 0.3 one spot to the right to get 3, then you must also move the decimal in 1.68 one spot to the right, getting 16.8, so the problem becomes 16.8 ÷ 3. Do long division to get the result:

$$
\begin{array}{r}
5.6 \\
3 \overline{)\ 16.8} \\
-\underline{15} \\
18 \\
-\underline{18} \\
0
\end{array}
$$

You can always multiply back to double check your result:

$$
0.3 \times 5.6 = 1.68
$$

Moving the decimal may seem confusing, but there are some easy shortcuts to remember.

- Moving the decimal to the *right* one space is the same as multiplying by 10; 2 spaces multiplies by 100; and so on, so the numbers get *bigger*.

- Moving the decimal to the *left* means you are dividing by 10 for each space moved, and the number always gets *smaller*.

- Always think about your answer to check that it is reasonable.

Moving the decimal to the right is the same as multiplying by 10 for each space moved. Moving it to the left is the same as dividing by 10 for each space moved. ∎

✔ **QUICK CHECK**

Solve these problems:

1. $1.27 + 3.6 =$ _____

2. $14.87 - 3.2 =$ _____

3. $2.4 \times 1.2 =$ _____

4. $8.4 \div 0.2 =$ _____

PERCENTS

As you learn anatomy and physiology, you will encounter many physiological values that are stated as percents. Percents are based on 100, where 100% is the total. For that reason, when working with percents, always be sure they add up to 100% and no more than that.

Percents are easy to work with. They are essentially fractions expressed as hundredths. For example, 25% is the same as 25/100, which can be further reduced to 1/4. And because fractions can be expressed as decimals, so can percents. You simply put the decimal two places to the left of the percent. So, 25% becomes 0.25 and 7% becomes 0.07. Percents, decimals, and fractions are all interchangeable, but when doing math operations, percents should be converted into either decimals or fractions. You cannot do math operations with mixed expressions— they must all be whole numbers, or fractions, or decimals. **Table 2.1** explains the relationship between these expressions.

TABLE 2.1   **The relationship between percents, decimals, and fractions.**

Percent	Decimal	Fraction in hundredths	Reduced fraction
10%	0.10	10/100	1/10
25%	0.25	25/100	1/4
40%	0.40	40/100	2/5
50%	0.50	50/100	1/2
75%	0.75	75/100	3/4
100%	1.00	100/100	1

Answers: 1. $1.27 + 3.60 = 4.87.$ **2.** $14.87 - 3.20 = 11.67.$ **3.** $2.4 \times 1.2 = 2.88.$ **4.** Move the decimal in the divisor and dividend both space one to the right, so it becomes $84 \div 2 = 42.$

TIME TO TRY

Can you supply the missing information in this table?

Percent	Decimal	Fraction
36%	_____	_____
_____	0.42	_____
_____	_____	80/100 = 4/5

Understanding Your ABCs: **Calculating Your Grade**

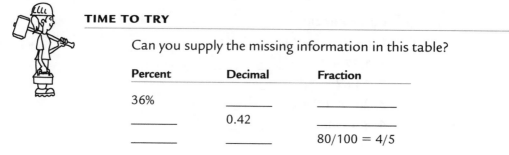

A responsibility of being a college student is tracking your grade. Record each grade you receive, including how many points you received and how many were possible. You're ultimately interested in how you do in the course, and *that* final grade is based on all the grades you receive. It's essential that you track both your individual grades and your overall course grade all semester. To do this, you need math. You also need to know how your course grade is determined. In the most common system, your grade is determined by the percentage of possible points you receive, but you need to know your instructor's grading scale. Many use a standard 10% scale:

90.0–100% = A   80.0–89.9% = B

70.0–79.9% = C   60.0–69.9% = D   < 60% = F.

Some instructors use variations of this with different cutoffs for the letter grades. Other instructors or institutions may designate "+" or "−" grades. You must know what scale will be used.

Let's try an example. Assume the grading scale is the standard 10% scale shown above. If you get 69 points correct on a test, what is your letter grade? _____ *Wait—not so fast!* Before determining your grade, you need to know how many points were possible. If it was a 100-point test and you got 69 out of 100 right, then you got 69% correct, which is a D. Sorry. But what if, instead, the test was worth 75 points? To calculate

Answers: 36% = 0.36 = 36/100 = 9/25; 42% = 42/100 = 42/100 = 0.42 = 42%; 80% = 21/50 = 80/100 = 0.8 = 4/5.

your grade, divide the number of points you got correct by how many points were possible. For this example, divide 69 correct by 75 possible, which gives you 0.92. To convert the decimal to a percent, multiply by 100. You get 92%, which is an A. Nice job! But if the exam was worth 120 points, you got 69 out of 120, which is only 57.5%, and you failed. Always divide the number of points you get correct by how many points were possible, and then multiply by 100 to get your percent.

If your course grade is also determined by the percent of points you receive during the semester, you can calculate your course grade at any time. **Table 2.2** provides an example. You see six grades listed in column 3, column 4 shows how many points each item was worth. If these were all of your grades, what letter grade would you have at this point in the course? Add all of your points together in column 3, then divide that number by the total of possible points in column 4. Simple math shows you that you've received 103 points out of a possible 125. Now divide your points by the total possible: 103/125 = 0.824, then multiply that by 100 to convert the decimal to a percent: $0.824 \times 100 = 82.4\%$. On the grading scale we're using, that means you have a B. Nice! Recalculate your course grade with each new grade you receive as the semester progresses. It's your responsibility to know your grade at all times. Don't wait for your instructor to let you know if you're not doing well—keep track yourself and work harder, or seek assistance immediately if you're not getting the grade you want or if your grade begins to drop.

TABLE 2.2 **Example of recorded grades.**

Date	Item	Points I Got	Possible Points
Jan 20	Quiz 1	17	20
Jan 27	Quiz 2	19	25
Jan 28	Lab 1	8	10
Feb 3	Test 1	43	50
Feb 5	Lab 2	4	5
Feb 6	Homework	12	15

Can You Feel the Power? **Understanding Exponents**

We briefly discussed exponents earlier, when we stated that a number written with an exponent is basically a multiplication problem: $2^3 = 2 \times 2 \times 2 = 8$. In science, very large and very small numbers are often written in a special format that uses exponents based on powers of 10. This format is called **scientific notation.** Let's consider the number 200 to see how this format is used. To write a number in scientific notation, place the decimal immediately after the first digit, and then drop the zeroes. This number is called the **coefficient.** In this example, the coefficient is 2. Next, count how many spaces you moved the decimal—two places to the left. Each of those spaces represents a power of 10, so two places means $10 \times 10 = 100$. In scientific notation, we would write 200 as 2×10^2. As you can see, this is $2 \times 10 \times 10 = 200$. Remember, the first number in scientific notation must be greater than 1 but less than 10. If a number is less than 1, the exponent is a negative power of 10. For example, 0.0004 would be 4×10^{-4} because the decimal was moved four spaces to the right. You will rarely need to do math operations with scientific notation, so we will skip those. You should, however, understand scientific notation so that you can understand some of the measurements you will read about—such as cell sizes, which are often measured in micrometers (10^{-6} meter). **Table 2.3** lists some common exponents.

TIME TO TRY

Express the two numbers below in scientific notation.

24,000,000 = _____

0.003 = _____

If you did this correctly, you got 2.4×10^7 and 3×10^{-3}.

TABLE 2.3   **The values of some common exponents used in scientific notation.**

Exponent	Value	Term
10^9	1,000,000,000	billions
10^6	1,000,000	millions
10^3	1,000	thousands
10^2	100	hundreds
10^1	10	tens
10^0	1	ones
10^{-1}	1/10	tenths
10^{-2}	1/100	hundredths
10^{-3}	1/1,000	thousandths
10^{-6}	1/1,000,000	millionths
10^{-9}	1/1,000,000,000	billionths

My Cell's Bigger than Your Cell! **Ratios and Proportions**

- Sodium and potassium move across cell membranes in a 3:2 relationship.

- In the United States, the ratio of males to females at birth is about 105:100.

- The ratio of males to females declines steadily until, after age 85, it is only 40.7:100.

HUH? Welcome to the comparatively interesting realm of ratios. A **ratio** expresses a relationship between two or more numbers—it is a way to compare them. Ratios can be expressed using a colon between the numbers (as above), as a fraction, or by using the word "to." For example, carbohydrates contain hydrogen and oxygen in a 2 to 1 ratio, meaning there are twice as many hydrogen atoms in carbohydrates as there are oxygen atoms.

Ratios are used for comparison, and they can also be expressed as fractions. For example, a ratio of 1:2 means the same as 1/2. Look at that carefully, though. Let's say the ratio of men to women in your anatomy class is 1:2. We're not saying that half of the class are men, we are saying there are half as many men as women.

What if you have a ratio and you want to know the fraction of the whole represented by each group? Let's say you are in a class where there are three men to one woman. The ratio of men to women would be 3:1. The "3" represents the men, and the "1" represents the women. Simply add the individual numbers in the ratio together to get the denominator—in this case that number is 4(3 + 1). Then use the individual numbers in the ratio as the numerators. That means that 3/4 of the class are men, and 1/4 are women.

If ratios can be expressed as fractions, they can also be expressed as decimals and percents. Because they can be written as fractions, they can also be reduced like fractions. For example, a ratio of 4:6 is the same as 4/6, which is the same as 2/3. When working with ratios, it is critical to write them in the correct order. If an anatomy class has 10 males and 20 females, the ratio of males to females is 10:20. If we write it as 20:10, it means there are twice as many males as females, which is not true.

TIME TO TRY

Empty your pocket or purse of change. Separate the coins by denomination. Count all of the coins in each category. Now express those numbers in a ratio: _____

Why is it important to indicate the order in which you are listing the coins? _____

_____

If you did this correctly, you should have indicated the order of the coins, because without that reference, we have no idea what number corresponds with which coin. Perhaps you had 5 pennies, 4 nickels, 3 dimes, and 2 quarters. If you wrote your ratio in that order, it would be 5:4:3:2.

We also use ratios to discuss quantities in a certain amount. For example, there are about 280 million hemoglobin molecules in each red blood cell. That can be expressed as 280 million/cell, which looks more like a fraction, but it is really saying the ratio is 280 million to 1. Rates are also a special type of ratio. A red blood cell travels about 700 miles in its 120-day lifespan, giving a ratio of 700 miles/120 days. Drug doses are also often given in this rate format; for example, 5 milligrams per kilogram of body weight.

Proportions are statements of equal ratios. A simple example would be to say 1/2 = 4/8. In science, we often use proportions to solve problems. To see how, examine a generic version:

$$\frac{a}{b} = \frac{c}{d}$$

Because these two ratios are equal, their cross-products are also equal, due to some basic laws of math. This means that the product of multiplying the first numerator (a) by the second denominator (d) equals the product of multiplying the second numerator (c) by the first denominator (b):

$$a \times d = c \times b$$

Let's say we want to know how many times the heart beats in an hour. Assume that the heart beats on average 80 beats per minute. We know there are 60 minutes per hour. So, we can set up the proportion, filling in the information we know and using "x" to represent the value we are trying to determine:

$$\frac{80 \text{ beats}}{1 \text{ minute}} = \frac{x \text{ beats}}{60 \text{ minutes}}$$

We know we can **cross-multiply** (**Figure 2.4**). When we do that, we get 4800 = x, so there are 4800 beats per 60 minutes, or per hour. In fact, there is a simpler way to write this problem, which is a shorter version of cross-multiplying. We know there are 80 beats per minute, and 60 minutes per hour, so we can calculate the beats per hour as follows:

$$\frac{80 \text{ beats}}{\text{minute}} \times \frac{60 \text{ minutes}}{\text{hour}} = \frac{4800 \text{ beats } \cancel{\text{minutes}}}{\cancel{\text{minute}} \text{ hour}} = \frac{4800 \text{ beats}}{\text{hour}}$$

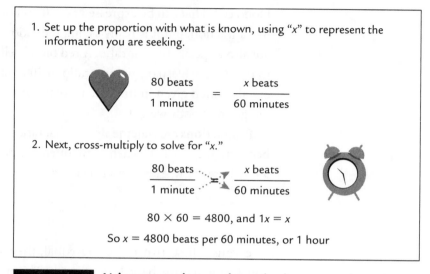

1. Set up the proportion with what is known, using "x" to represent the information you are seeking.

$$\frac{80 \text{ beats}}{1 \text{ minute}} = \frac{x \text{ beats}}{60 \text{ minutes}}$$

2. Next, cross-multiply to solve for "x."

$$\frac{80 \text{ beats}}{1 \text{ minute}} = \frac{x \text{ beats}}{60 \text{ minutes}}$$

$$80 \times 60 = 4800, \text{ and } 1x = x$$

So $x = 4800$ beats per 60 minutes, or 1 hour

FIGURE 2.4 **Using proportions to determine how many times the heart beats in an hour.**

Notice that the units are shown, and in the next to last step, minutes appear on both the top and bottom. That means they cancel each other out, so we are left with beats per hour, which is the correct unit. Using the units can be an easy way to ensure that you have set up the problem correctly. This is another example of taking the time to think through the problem before you start—the units should make sense when you are done.

You should be aware of another way in which proportions are used in science. You'll likely encounter the terms **directly proportional** and **inversely proportional.** These terms explain relationships. Often, but not always, they're used to compare rates of change. Consider breathing. To inhale, muscles contract and expand the size of your chest. Your lungs move with your chest wall, thus as your chest expands, so do your lungs, increasing the space inside them—the lung volume. The increase in lung volume is *directly proportional* to the expansion of your chest—they change together at a similar rate. Now let's look inside your lungs. The air inside exerts pressure on the walls of your lungs. As your lungs expand, there's more room for the air so it exerts less pressure. Thus, as lung volume increases, pressure inside your lungs decreases. They change at a similar rate but in opposite directions—as volume goes up, pressure goes down. Thus, lung volume and lung pressure are *inversely proportional.*

✔ **QUICK CHECK**

An average person takes about 12 breaths per minute. How many breaths do they take in an hour? _____

Who Ever Heard of a Centimeter Worm? **The Metric System**

In the United States, we all grew up learning there are 12 inches to a foot, 3 feet to a yard, and 100 yards to a football field, and we measure driving distances in miles, which contain 5280 feet. In the kitchen, we use cups, half cups, thirds of cups, quarter cups, tablespoons, teaspoons, eighths of teaspoons, pints, quarts, gallons, ounces, and pounds. There are so many units in our system that it is amazing we can keep them straight. What if I told you there is a simpler way to measure?

It is called the **metric system,** or **System Internationale (SI).** It is universally used in science and by almost every country in the world except the United States. You have undoubtedly had brushes with learning the metric system, and you may have found it difficult. The problem is not with the metric system, which is delightfully simple. Instead, the problem is with our complicated U.S. (also called English) system and the need to convert between the two systems. This requires—you guessed it—math.

WHY SHOULD I CARE?

Science uses metric measurement almost exclusively, so you will need a basic understanding of metric units for your coursework and for your future career. In addition, almost everyone on our planet—except the United States—uses the metric system.

Answer: Set up the proportion: 12 breaths/minute = x breaths/60 minutes. Cross-multiply: 12 × 60 = 720 = x, so x = 720 breaths per hour.

The metric system is amazingly simple because it is all based on the number 10, so obviously decimals are easier to use than they are in the U.S. system. For our purposes, we'll learn four main units used in science: those that deal with length or distance, mass, volume, and temperature. Each of these has a standard or base unit:

- The basic unit of length (or distance) is the **meter (m).**

- The basic unit of mass is, technically, the **kilogram (kg),** but many sources use the **gram (g)** as the base unit instead.

- The basic unit of volume is the **liter (L).**

- The basic unit of temperature is the **degree Celsius (°C).**

These are the base units, but more convenient units are derived from these. For example, a meter is just a bit longer than 3 feet (39.34 inches), so it is not a convenient unit for measuring the size of, say, your finger or a cell. Smaller units of the meter, based on the powers of 10, are used instead. These units are named by adding the appropriate prefix to the term *meter* (**Table 2.4**). **Centi-** means 1/100, and there are 2.54 centimeters (cm) in an inch, so centimeters work well for measuring fingers.

TABLE 2.4 **Metric system prefixes.**

Prefix	Symbol	Decimal equivalent (multiple)	Exponential equivalent (Scientific notation)
Pico-	p	0.000000000001	10^{-12}
Nano-	n	0.000000001	10^{-9}
Micro-	μ	0.000001	10^{-6}
Milli-	m	0.001	10^{-3}
Centi-	c	0.01	10^{-2}
Deci-	d	0.1	10^{-1}
no prefix		1.0	10^{0}
Deka-	D	10.	10^{1}
Hecto-	H	100.	10^{2}
Kilo-	k	1000.	10^{3}
Mega-	M	1,000,000.	10^{6}
Giga-	G	1,000,000,000.	10^{9}

Cells are microscopic, so they are best measured in even smaller units, such as *micro*meters—one micrometer = 1 millionth of a meter. Driving between cities, you can best measure the long distances in *kilo*meters, each of which equals 1000 meters. Again, all metric units are based on 10. Think about that—you first learn to count from 1 to 10, then you can count by tens to 100, then by hundreds to 1000, and so on. It is an easy system.

Table 2.4 provides many of the prefixes and their base-10 equivalents. In anatomy and physiology, you will use some units more than others. For length or distance, which is a straight linear measurement, you will mostly work in meters, centimeters, millimeters (1/1000 m), and micrometers. For mass, which is the actual physical amount of something, you will most often refer to kilograms (1000 grams), grams, and milligrams (1/1000 g). For volume, which refers to the amount of space something occupies, the most common units will be liters and milliliters.

The Celsius temperature scale, also sometimes referred to as the centigrade scale, does not use prefixes. Instead, it has a single unit: degrees Celsius. However, the scale is different than the Fahrenheit scale with which you are familiar. Water freezes at 0°C and boils at 100°C (**Figure 2.5**).

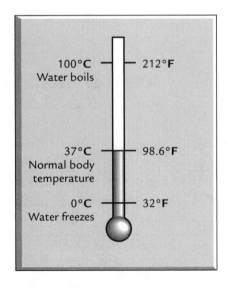

FIGURE 2.5   **Celsius and Fahrenheit temperature comparison.**

At what Fahrenheit temperature does water freeze? _____
Boil? _____ In the Celsius scale, normal body temperature (98.6°F) is 37°C. To convert between the two temperature scales, there are two specific equations—one to convert from degrees Celsius to degrees Fahrenheit, and another to convert in the opposite direction. Each of these equations is listed below, first in its original form, which includes a fraction, and then with the fraction converted to a decimal. You will likely use a calculator to do any conversions and it will be easier to multiply using the decimal.

$$°Celsius = (°Fahrenheit - 32) \times 5/9$$
$$= (°Fahrenheit - 32) \times 0.556$$
$$°Fahrenheit = °Celsius \times 9/5 + 32$$
$$= °Celsius \times 1.8 + 32$$

Let's try one of each type of conversion. To convert 37°C to °F, we use the second equation:

$$°F = 37°C \times 1.8 + 32$$
$$= 66.6 + 32$$
$$= 98.6°F$$

Now let's convert 75 °F to °C. We use the first equation as follows:

$$°C = (75°F - 32) \times 0.556$$
$$= 43 \times 0.556$$
$$= 23.9°C$$

Most of the work you will do, though, will be converting units for length, mass, and volume, so let's move on.

PICTURE THIS

If you drink carbonated soft drinks, you are likely quite familiar with their standard large plastic bottles. What is their volume in metric units? _____ liters

Knowing this, envision the bottle only half full. Is that amount more or less than a gallon of milk? _____

Answers to Picture This: A meter is longer than a yard by a few inches. A pound contains just over 450 grams.

TIME TO TRY

You will use the metric system more easily if you have some idea of the size of the base units. Most packaged items manufactured in the United States list both metric and U.S. units. Most rulers and measuring tapes have both metric and U.S. units. Explore your home and see if you can determine the following:

■ Which is longer, a meter or a yard? _____

■ Which is heavier, a gram or a pound? _____

Know the paperclip! A standard small paperclip has a mass of about 1 gram (it's very light). A standard large paperclip has a side-to-side width of about 1 cm, and the wire from which it is made has a diameter of about 1 mm. ■

You will become more familiar with the metric system as you use it. You may occasionally need to convert from U.S. units to metric units, although this is done more as an exercise than out of need—in class almost all measuring and discussion will use metric units. Still, it is useful to know how to convert between the two systems. **List 2.1** on page 78 shows some of the basic conversion factors. Please note the first relationship in the list—it's very important in the medical field. When measuring liquids, 1 milliliter (mL) equals 1 cubic centimeter (cc), so liquid medications may be prescribed or administered with either unit—mL or cc.

Answers to Time to Try: 2 liters; Half of a 2-L soda bottle is definitely less than a gallon of milk, so a liter is smaller than a gallon.

Let's try some conversions. First, let's do the easy stuff: converting between metric units. Remember that in the metric system the difference between the units will always be some multiple of 10. Let's convert 13 meters into centimeters.

$$13 \text{ m} = \underline{\hspace{3cm}} \text{ cm}$$

A centimeter is 1/100 of a meter, so there are 100 centimeters per meter. Thus:

$$13 \text{ m} \times 100 \text{ cm/m} = 1300 \text{ cm}$$

Now we will convert 27 millimeters into centimeters, but let's try another method. All we really have to do to convert between different metric units is move the decimal, but by how many spaces and in which direction? How many spaces you move is determined by the difference in the power of 10. We know that millimeters are thousandths of a meter, and centimeters are hundredths of a meter.

$$\text{millimeters} = 10^{-3} \qquad \text{centimeters} = 10^{-2}$$

LIST 2.1   **Some basic conversions.**

SPECIAL RELATIONSHIPS
1 milliliter (mL) = 1 cubic centimeter (cc), a unit often used in administering liquid medications
The mass of 1 milliliter of water = about 1 gram
The mass of 1 liter of water = about 1 kilogram

APPROXIMATE CONVERSION FACTORS
Multiply **inches** $\times$ 2.54 cm/inch to get **centimeters**.
Multiply **feet** $\times$ 0.305 m/foot to get **meters**.
Multiply **miles** $\times$ 1.6 km/mile to get **kilometers**.
Divide **pounds** by 2.2 pounds/kg to get **kilograms**.
Multiply **gallons** $\times$ 3.8 L/gallon to get **liters**.

$^{\circ}$Celsius = ($^{\circ}$Fahrenheit -32) $\times$ 5/9 = ($^{\circ}$Fahrenheit -32) $\times$ 0.556
$^{\circ}$Fahrenheit = $^{\circ}$Celsius $\times$ 9/5 + 32 = $^{\circ}$Celsius $\times$ 1.8 + 32

So, if we look at the exponents, they are different by one. We will move the decimal in our number (27) by one spot. But in which direction? When converting from smaller to larger units, the decimal moves left. When converting from larger to smaller units, the decimal moves right. Back to our example: converting 27 mm to cm gives us 2.7 cm.

Look at **Figure 2.6** to get a better understanding of how far and in what direction you need to move the decimal when doing these conversions.

When converting within metric units:

1. Put the units in scientific notation and subtract the smaller exponent from the larger one. The difference is how many spaces the decimal will move in your coefficient.

2. If you are converting from small units to larger ones, the number gets smaller, so the decimal moves to the left. If you are converting from larger units to smaller ones, the number gets bigger, so the decimal moves to the right. ■

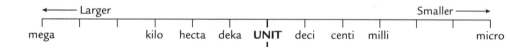

FIGURE 2.6 **Converting from one metric unit to another.** Let's say you are converting meters, the base unit, into centimeters. This illustration shows that to get from meters to centimeters, which are smaller, you would move two spaces to the right. That means you need to move the decimal two places to the right. If you are converting meters to kilometers, which are larger, you would move the decimal three spaces to the left.

TIME TO TRY

Now that you see the simple secret to this process, complete the following conversions:

5 kg = _____ g   8 mL = _____ L   6 cm = _____ m

If you did these conversions correctly, you should see that 5 kg = 5000 g; 8 mL = 0.008 L; 6 cm = 0.06 m. See, the metric system is easy!

Now let's convert from metric to U.S. units, and vice versa. To do these you need the correct conversion factor (see List 2.1). Let's convert 18 inches into centimeters. From the list we see that there are 2.54 cm per inch, so we merely multiply 18 inches × 2.54 cm/inch. The answer is 45.72 cm.

Let's convert 30 cm into inches. We can try a proportion to solve this.

$$\frac{2.54 \text{ cm}}{1 \text{ inch}} = \frac{30 \text{ cm}}{x \text{ inches}}$$

Cross-multiplying gives us $2.54x = 30$, so to get x we divide both sides by 2.54:

$$30 \div 2.54 = 11.81 \text{ inches}$$

You now have all the tools you need to do conversions between the two systems. The easiest solution, however, is to only work in metrics, like the rest of the world!

✔ **QUICK CHECK**

Complete these conversions:

1. 30 miles = _____ km
2. 8 L = _____ mL
3. 110 pounds = _____ kg

How Do You Measure Up? **Basic Measurement**

Now that you understand the basic units of measurement, you need to know how to measure. A common error in scientific experimentation is called human error, which includes math mistakes (which you won't make now!) and something as simple as not measuring correctly. When you're baking brownies, adding extra sugar and chocolate may be a good thing, but that won't work in science. Measurements must be done precisely and with appropriate equipment.

3. 110 pounds ÷ 2.2 pounds/kg = 50 kg.

Answers: 1. 30 mi × 1.6 km/mi = 48 km. 2. 8 L × 1000 mL/L = 8000 mL.

MEASURING LENGTH

Length is usually measured with a meter stick or a ruler. Grab one—surely you can locate one somewhere. I'll wait . . .

Examine the scale. There are likely two scales—inches and metric. On a meter stick, the scales are often on opposite sides of the stick. Look at the metric scale. If it is a meter stick or metric tape measure, the small numbered units are usually centimeters. Confirm that there are 100 of these in a meter. Compare the size of a centimeter to an inch by placing your fingers on each side of a centimeter, then maintaining that space as you move to the inch scale. Note that an inch is a bit over 2.5 times bigger—2.54 to be exact (**Figure 2.7a**).

Now examine the space between 0 and 1 cm. Count the spaces. How many are there? _____ You should have counted 10 spaces. These tiny units are millimeters, each equal to 1/1000 meter. That makes sense—

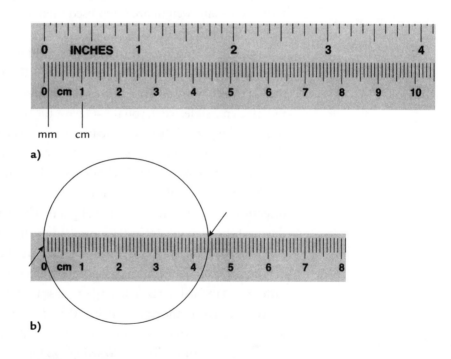

FIGURE 2.7   **Using a metric ruler. a)** This shows the comparison of the metric and inch scales. Millimeters and centimeters are labeled. **b)** Always start measuring at the very edge of the object, positioning the ruler carefully so that the scale runs precisely through the dimension being measured.

each millimeter is 1/10 of a centimeter, which is 1/100 of a meter, and 1/10 of 1/100 is, indeed, 1/1000.

When measuring with a ruler, be sure to align the zero line exactly at the edge of your object and measure exactly to the far edge. Do not round off. If you are measuring the diameter of a circle, be sure your ruler is positioned across the widest part of the circle (**Figure 2.7b**).

When measuring length, be sure you use the appropriate scale. ■

MEASURING MASS

Mass is the actual amount of something, and it is closely related to weight, but weight takes into account the force of gravity acting on the mass. Mass will be constant, but weight will vary with gravity—just ask the astronauts, who have no weight in outer space where there is no gravity. Here on Earth, where the gravitational pull is rather steady, the terms mass and weight are often used interchangeably, but you should realize that they are different.

Mass can be measured with a digital scale that measures in grams, but often in labs we use the triple beam balance (**Figure 2.8a**). Before using the balance, be sure all the attached standard masses are pushed as far to the left as possible. Next, you must "zero" the balance before adding anything to the pan. Notice how the two lines on the far right line up. One line is on the arm and moves with it; the other is on the end piece of the scale. If they are not perfectly aligned, slowly turn the zero knob located on the left of the scale, usually under the pan. Rotate the knob in either direction until the two lines are aligned. The arm will move up and down a bit—wait until it has stopped and the lines are aligned. The scale is now zeroed. NOW you are ready to measure the mass of your object.

There are three beams on the arm of the balance, each suspending a different standard mass. The largest mass is 100 g. Each spot that you move that mass to the right equals 100 g. Another beam has a 10-g mass, and the front beam has a 1-g mass. Starting with the 100-g mass, slide it across until it causes the arm to swing too far to the right. That was too heavy, so back it up to the left by one spot. Note the number—that is how many hundred grams are in your object. Next, slide the 10-g mass until it is too much, back up one notch, then note how many 10's of grams there are. Finally, slide the 1-g mass carefully until the two lines on the right side again align, as they did at the beginning. At that point, the scale is balanced. Read all of the whole numbers on the scale; each line beyond the last whole number is 1/10 g. If the 100-g mass is at 200, the 10-g mass is at 80, and the 1-g mass is halfway between the 3 and 4, what is the mass of the object? _____

(It would be 283.5 g, as shown in **Figure 2.8b.**)

Always zero the balance before starting, with all standard masses at the far left. ■

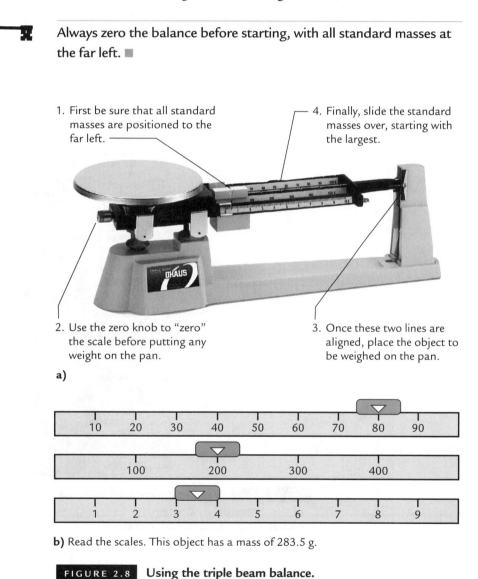

1. First be sure that all standard masses are positioned to the far left.

4. Finally, slide the standard masses over, starting with the largest.

2. Use the zero knob to "zero" the scale before putting any weight on the pan.

3. Once these two lines are aligned, place the object to be weighed on the pan.

a)

b) Read the scales. This object has a mass of 283.5 g.

FIGURE 2.8   **Using the triple beam balance.**

MEASURING VOLUME

Volume refers to the amount of space a substance occupies. In anatomy labs, you will most often measure liquid volumes by using beakers or graduated cylinders. Always use the smallest container in which the substance will fit—the smaller it is, the more accurate your reading will be. Always read the scale, usually in milliliters, at eye level for accuracy. You should also know how to read the meniscus (**Figure 2.9**)—this is especially critical in a graduated cylinder. Liquid in a container tends to climb slightly up the sides, so the center is lower than the edges, where the liquid contacts the container. This dip is called the **meniscus.** When reading the scale, always read it at the low point of the meniscus for the best accuracy.

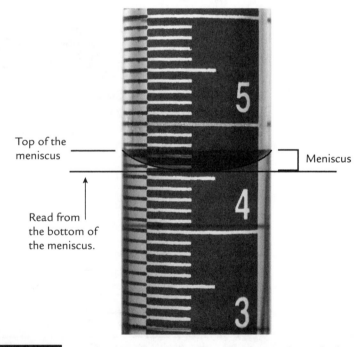

Top of the meniscus

Meniscus

Read from the bottom of the meniscus.

FIGURE 2.9 **Read measurements of liquid volumes from the bottom of the meniscus.** By doing so, you see that this graduated cylinder contains 4.55 mL of liquid, not the 4.75 indicated by the top of the meniscus.

TIME TO TRY

Find the narrowest clear container you can and fill it halfway with water. Look at it at eye level. Use a ruler on the outside of the container to measure the highest point of the water. _____ Now measure the lowest point. _____ The dip that you see is the meniscus. Whenever you measure a liquid volume, always measure at the lowest point of the meniscus.

You Ought to be in Pictures: **Tables, Graphs, and Charts**

We explored how to get numbers by measuring and how to work with them. Now we will see how these numbers and other information, collectively called data, can be depicted.

TABLES

We have already used tables in this book, and you should be familiar with them, so let's quickly review the basics. Tables come in many forms and are a convenient way to present information so it is easy to read and compare. We will use **Table 2.5** as a reference.

When viewing a table, start with the table title, in this case "Fluctuations in body temperature throughout the day." The title usually tells you what the table contains. Next, realize that tables are carefully arranged in columns and rows. All information in a single column is related, and all information in a single row is related. Look at the top of each column—these are column heads that tell you what information

TABLE 2.5   **Fluctuations in body temperature throughout the day.**

Subject	T °F at 4 AM	T °F at 8 AM	T °F at Noon	T °F at 4 PM	T °F at 8 PM	T °F at Midnight
A	96.0	96.3	96 4	96.8	97.0	96.8
B	98.0	98.4	98.6	98.8	99.0	98.4
C	97.6	98.0	98.4	98.5	98.9	98.6
D	97.2	97.3	97.5	97.6	98.0	97.6

each column contains. Look at the beginning of each row. These are row heads, or labels that tell you what each row contains. In our example, the column heads reveal that the first column identifies each test subject, and the other columns contain the temperatures for all test subjects at specific times of the day. The row heads tell us that all temperatures in a single row belong to a single test subject, and who it is. So, by using the column and row heads, it is easy to find out, for example, what temperature Subject C had at noon.

TIME TO TRY

On a separate piece of paper, using sentences and paragraphs, write out all the information that is included in Table 2.5. Which version is easier to read? In which version can you more easily determine Subject D's temperature at 8 PM? _____

GRAPHS

Graphs present a more pictorial view of data. Numerical data that can be organized in a table (**Figure 2.10a**) can usually also be presented in a graph. The main advantage is that the graph allows you to spot trends and relationships almost instantly. There are various types of graphs. Let's look at three of them: line graphs, bar graphs, and pie charts.

LINE GRAPHS   Look at **Figure 2.10b**. This is a line graph, and they usually are laid out in a grid. The horizontal axis at the bottom is the *x*-axis. It often, but not always, marks the progression of time. The vertical axis on the left is the *y*-axis and it typically reflects some increasing value. Where these lines meet on the lower left of the graph marks the 0 position, so units go up as you move away from that point. Each axis should be clearly labeled and should include units so the viewer can easily understand.

In our example, the *x*-axis tells us who the data are about—the age group of children for whom the listed height is typical. Note that the units on the *x*-axis include both months and years. If the ages were all years, the units (years) would be listed only in the *x*-axis label line, as they

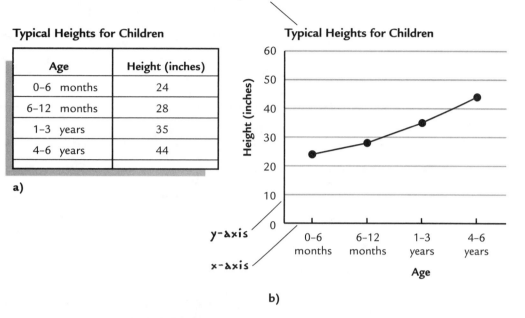

FIGURE 2.10 **Typical heights for children through age 6 years.**
a) Table format. **b)** Basic line graph.

are on the *y*-axis, instead of under each bar. The *y*-axis gives height in inches. For each age group, the height (data point) is placed above the age it represents. The data points may be left unconnected, they may be directly connected (as they are here), or a line of "best fit" may be drawn that passes between the points so they are evenly distributed on each side of it. In our example, because this depicts growth with time, the line allows us to see at once the heights for each age group and to quickly comprehend the trend of a gradual increase in height with age.

When drawing a line graph, don't forget to label the axes and to include units. The most common mistake made when graphing data is to not use an appropriate scale (**Figure 2.11**). Be sure to size the units to maximize the space the graph fills. You don't want the graph to be cramped into one corner, making it hard to read. Spread it out both vertically and horizontally.

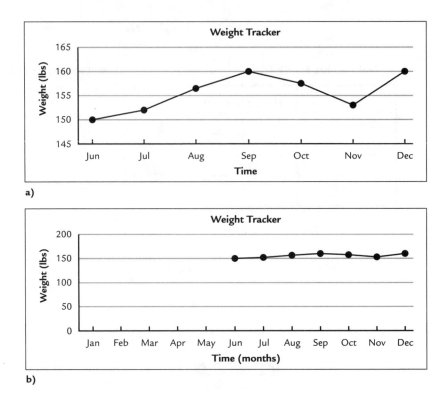

a)

b)

FIGURE 2.11   **Comparing line graphs. a)** This graph is poorly done. Notice that the scale is too large, so the data are crunched together and their values are hard to read. The available space is poorly used. **b)** This graph shows the same data, but the scale has been adjusted so the data fill the space better and the individual values are easier to read.

BAR GRAPHS   **Figure 2.12** is a bar graph showing the amount of water in the human body. Looking at the axes, you see that the *x*-axis has three separate categories: Total water, intracellular fluid (the water located inside body cells), and extracellular fluid (the water not contained in cells). Each of these categories has two bars—one for males and one for females. The *y*-axis tells what percent of the total body weight the water represents. This graph is drawn so that the male and female data are directly compared by being positioned side-by-side, yet they are easily distinguished by use of different shading. The shading is explained on the lower left in a feature called the **key.** This bar graph has a 3-D effect that does not change its meaning at all—it just makes it slightly

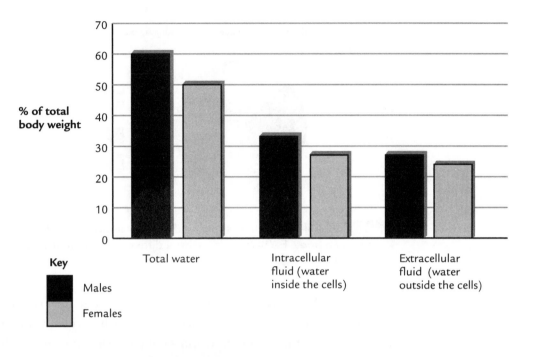

Key

Males

Females

FIGURE 2.12 **A simple bar graph showing a comparison of body water in males and females.**

more interesting. Bar graphs may be drawn vertically or horizontally. Because each bar is so distinct, these graphs are good for comparing specific bits of data, whereas line graphs may be better at showing an overall trend.

PIE CHARTS **Figure 2.13** is a pie chart, a type of graph that shows parts of a whole. This one shows the main molecules that make up the human body and tells the percent of the body made of each type of molecule. When looking at this, you immediately see that all the parts add up to the whole "pie," which is 100%, so these charts are very effective when showing percents. We all have a visual concept of a whole pie and a slice of pie, so even before looking at the numbers, we instantly see the differences in distribution. You know right away from this pie chart that one type of molecule makes up well over half of the body.

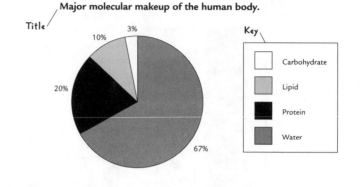

A pie chart showing the major molecular makeup of the human body. Each slice represents a different type of molecule, as indicated in the key.

Pie charts have no axes to label and the space within them is limited, so a key is often used. Labels for each "slice" may be written within the pie or placed to the outside, as in our example. Using different shading or colors for the slices makes these graphs even more readable. Because there are no axes to provide information, both the graph title and key are very important.

When reading a graph, always read the title first, then the axes, key, and labels. Finally, just let your eyes take in the relationships that are depicted. ▪

✔ **QUICK CHECK**

1. In Figure 2.10, which age group is shortest? _____
What is the typical height for children aged 1 to 3 years?

_____

2. In Figure 2.12, what percent of the total body weight is water in an average female? _____
3. In Figure 2.13, which molecule makes up most of the human body? _____

Answers: 1. The shortest group is aged 0–6 months, and children aged 1–3 are typically 35 inches tall. 2. 50% 3. water

Final Stretch!

Now that you have finished reading this chapter, it is time to stretch your brain a bit and check how much you learned. For online tests, tutorials, animations, activities, web links, and an ebook, visit the *Get Ready for A&P* companion website.

RUNNING WORDS

At the end of each chapter, be sure you have learned the language. Here are the terms introduced in this chapter with which you should be familiar. Write them in a note-book or enter them in your computer. Define them in your own words, then go back through the chapter to check your meaning, correcting as needed. Also try to list examples when appropriate.

Data
Product
Exponent
Dividend
Divisor
Quotient
Normal
Average
Mean
Numerator
Denominator
Equivalent fraction
Reduce
Greatest common
    factor (GCF)
Common
    denominator
Least common
    multiple (LCM)
Repeating decimal
Scientific notation
Coefficient

Ratio
Proportion
Cross-multiply
Directly proportional
Inversely
    proportional
Metric system (SI)
Meter (m)
Kilogram (kg)
Gram (g)
Liter (L)
Degree Celsius (°C)
Centi-
Milli-
Kilo-
Mass
Volume
Meniscus
x-axis
y-axis
Key

WHAT DID YOU LEARN?

Try these exercises from memory first, then go back and check your answers, looking up any items that you want to review. Answers to these questions are at the end of the book.

PART A: SOLVE THESE PROBLEMS.

1. $(4 \div 2) + 6 - 5 \times 2^3 = $ _____

2. $2 \times 10^4 = $ _____

3. 27/36 reduced is _____

4. $3/8 \times 2/3 = $ _____

5. $5/6 - 7/12 = $ _____

6. $0.5 \times 0.4 = $ _____

7. If the respiratory rate is 12 breaths per minute, how many breaths are taken in 1 hour? _____

8. 4 meters = _____ centimeters

9. 1 inch = 2.54 cm, so 1 foot = _____ cm

10. If a man weighs 200 pounds and 60% of his body weight is water, how many pounds of water does he have? _____

PART B: ANSWER THESE QUESTIONS.

1. What is the mean of 8, 9, 12, 18, and 23? _____

2. What is the numerator in 4/5? _____

3. Express 3/10 as a decimal _____ and as a percent _____

4. From Figure 2.13, how much of the human body is made of protein? _____

5. What would you be measuring if you are looking at a meniscus? _____

6. In the metric system, list the base unit for each of the following:
   mass: _____
   length: _____
   volume: _____

7. What is meant by "normal" blood pressure? _____ _____

8. At what Celsius temperature does water boil? _____

9. The amount of space something occupies is called _____.

10. How many milligrams are there in 1 gram? _____

3 Terminology

You Say Humerus, I Say Funny Bone

When you complete this chapter, you should be able to:

▨ Break down anatomical and medical terms to understand their meanings.

▨ Build anatomical and medical terms from basic roots, prefixes, and suffixes.

▨ Describe the anatomical position and understand its relevance.

▨ Correctly use basic anatomical terms to describe the relative positions of body parts, planes and sections, and general body regions.

Your Starting Point

Answer the following questions to assess your knowledge of anatomical terminology.

1. Most anatomical terms arise from which languages? _____

_____

2. What is the difference between *anatomy* and *physiology*? _____

_____

3. What is meant by the term *cardiomyopathy*? _____

_____

4. What is meant by *the anatomical position*? _____

_____

5. What is meant by the term *medial*? _____

_____

6. What plane passes through the body from front to back? _____

How Do You Say, In Your **Language** . . .?

Have you ever read a computer manual only to find that you are still unsure of how to configure your personal firewall, or for that matter why you should? Has your mind gone numb as an auto mechanic explained all those expensive malfunctioning parts that were replaced to get rid of that mystery noise in your car? Have you tried to read the small print in an advertisement for a new prescription drug? At such times, it may seem as if other people speak a secret language. Indeed,

Answers: 1. Latin and Greek. 2. Anatomy studies the body's structures; physiology studies their functions. 3. Disease of the muscle of the heart. 4. The body position used as a common reference point—standing erect, everything facing forward, including palms. 5. Toward the midline. 6. Sagittal.

most professions have their own sublanguage, as do most academic disciplines. Anatomy and physiology certainly are not exceptions.

Learning the terminology in these two disciplines may be a challenge, but I won't say it is difficult. *Difficult* leaves the option of ducking behind excuses. I have often heard my own students say, "I could do this if the words weren't so hard, but this is just too difficult for me!" Instead, you should think of the words as a challenge, because a challenge sparks the competitor in us—it makes us try harder. Although the words may seem intimidating at first, you will quickly learn tools and tricks to help you understand even the longest terms. Are YOU up for the challenge?

PICTURE THIS

Upon arriving at your long-anticipated vacation destination, you notice how wonderfully new and exotic everything seems—the smells in the air, the people's faces and clothing, the food sold from vendors' carts, and the sounds. You are in paradise! Then you discover that both your luggage and your wallet are missing. You approach an authority and start explaining your situation only to be met with a blank stare. You suddenly realize part of the exotic sounds you have been hearing are a very different spoken language.

1. How will you convey your plight to this person?_____

_____

2. How difficult would it be if you could speak the local language?

_____

3. What could you have done before your trip to avoid this communication gap? _____

_____

Obviously if you plan to spend time in a place where another language is spoken, communication may be a challenge unless you learn the local language in advance. Your anatomy and physiology class will be a place where another language is spoken. Much of the terminology will be new, and most of the words do, in fact, come from languages other than English. You will learn about body parts and their functions by listening to lectures filled with this new language, and you will be expected to discuss your course material and write exam answers using this new language. That is why it is so important that you learn the language of anatomy and physiology. As you study, the first thing you should do is master—not just read, but *master*—the new words. You have to know the language before you can understand and join in the conversation.

Learning the vocabulary is the first step in learning A&P. ■

I recommend to my students that they maintain a running vocabulary list. The easiest way to do this is to keep a separate notebook into which you write all new terms as they are introduced (**Figure 3.1**). If you write new terms in your notes during lecture, be sure to transfer them to your vocabulary list and check their spellings later. Once you know that a term is spelled correctly, write its actual definition—exactly what does the word mean? You can find the actual textbook definition in your textbook or in a medical dictionary, but you should also try to explain the term in your own words.

Next, be sure you can use the word properly in a sentence. Try to add some examples that illustrate the term, if appropriate. For example, a tissue is defined as a group of cells organized together that share a common function. Examples of tissue include, bone, blood, cartilage, fat, and muscle. If you are a visual learner, try illustrating your new words if you can. If you are an auditory learner, try reading the words and their meanings out loud and consider tape recording your list. If you are a tactile learner, you might benefit from typing your list into a word-processing file, then alphabetizing it. You might also write the terms and their meanings on flashcards.

You may be surprised at the beginning of the course at the number of new terms you encounter. You can think of the words as a kind of smoke screen—the underlying principles of anatomy and physiology are rather simple, but you may not see that through all the smoke. More than one student has jokingly commented to me that "It's all Greek to me." They don't realize that their joke is not far from the truth—most anatomical terms have either Greek or Latin origins, although many

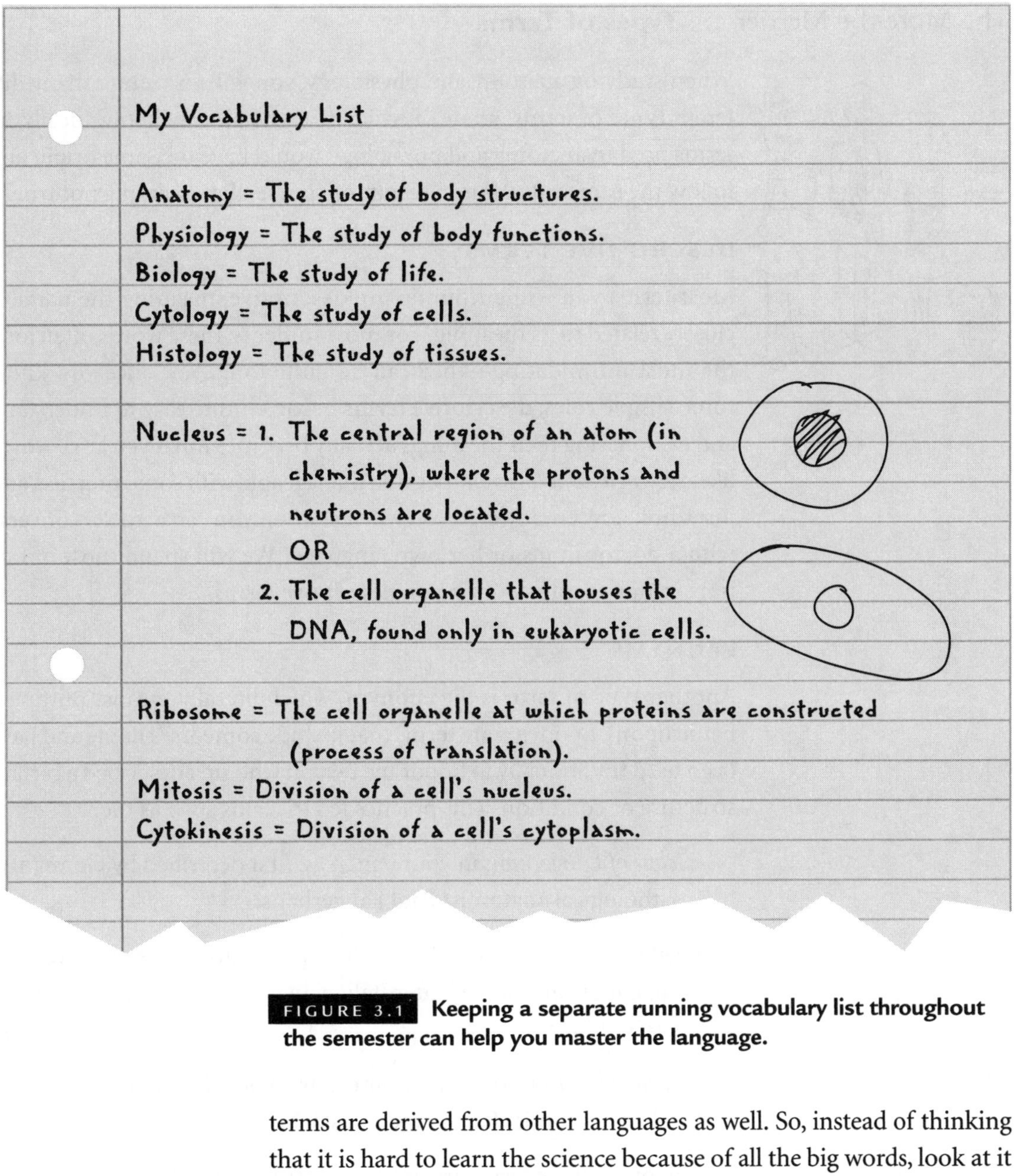

My Vocabulary List

Anatomy = The study of body structures.
Physiology = The study of body functions.
Biology = The study of life.
Cytology = The study of cells.
Histology = The study of tissues.

Nucleus = 1. The central region of an atom (in chemistry), where the protons and neutrons are located.
OR
2. The cell organelle that houses the DNA, found only in eukaryotic cells.

Ribosome = The cell organelle at which proteins are constructed (process of translation).
Mitosis = Division of a cell's nucleus.
Cytokinesis = Division of a cell's cytoplasm.

FIGURE 3.1 **Keeping a separate running vocabulary list throughout the semester can help you master the language.**

terms are derived from other languages as well. So, instead of thinking that it is hard to learn the science because of all the big words, look at it this way: While learning the science, you will learn some new languages!

✔ **QUICK CHECK**

From which languages do most anatomical terms originate?

Answer: Greek and Latin.

The More the Merrier . . . **Types of Terms**

When studying anatomy and physiology, you will encounter many different types of terms, many of which you will already know. Ideally, all terms used in anatomy and physiology would have the same origin and follow the same rules, but in reality, of course, that is simply not true.

DESCRIPTIVE TERMS

Most terms you will encounter are descriptive, meaning the name is closely related to its meaning. For most students, these words often look the most intimidating—they can be quite long. Yet, once you know some simple rules, **descriptive terms** become quite easy to understand and deciphering their meanings actually becomes fun. Even large words like *electroencephalography* become manageable with some practice, and just think how empowered you will feel when you can actually converse with a doctor in his or her own language! We will spend most of our time in this chapter working on descriptive terms.

EPONYMS

Another type of term is the **eponym,** which literally means "putting a name upon." Eponyms are terms that include someone's name and have been used traditionally to honor the person who first described a certain structure or condition. This practice led to terms such as the:

- *Islets of Langerhans* in your pancreas, first described by German pathological anatomist Paul Langerhans;

- *Sphincter of Oddi,* the "door" where pancreatic juice enters your small intestine, named after Italian anatomist and surgeon Ruggero Oddi; and

- *Eustachian tubes* between your ear and your throat, named after Italian anatomist Bartolomeo Eustachi.

Obviously, these terms are not so easily understood. Few people now associate an eponym with the discoverer, making the terms harder to master. Recent practice in anatomy has moved away from eponyms to a preference for descriptive terms, so those will be our focus. You will learn relevant eponyms as you move through your course, but fewer and fewer are in use today.

Of course, there are exceptions. Down syndrome was previously referred to as *mongolism* because the characteristic eye appearance of someone with Down syndrome was a bit like that of people of Mongolian descent. This term, however, is deemed offensive, so the eponym is now used more often. Another descriptive term for this genetic disorder is trisomy 21, which tells us that three copies of chromosome 21 cause this condition. Although few people know what trisomy 21 means, most people recognize and understand the eponym.

Most of the terms for your course are part of a broader area known as *medical terminology*. Eponyms are still used relatively commonly in medical terminology and in health-related fields, especially for naming diseases or abnormal conditions. You will occasionally encounter them, and you really just need to memorize them as they come up.

TIME TO TRY

Parkinson's disease is also called *paralysis agitans*.

1. Which is the descriptive term? _____

2. Which is the eponym? _____

3. With which term are you more familiar? _____

4. The first vertebra of your spine, located just under your skull, is also called the *atlas*, which is an eponym. Either recall who Atlas was in Greek mythology, or look up his name. How is this eponym also a descriptive term? _____

_____

Although paralysis agitans is the descriptive term, you are likely more familiar with the eponym—Parkinson's disease. You also likely realize that the atlas (bone) holds up your head, similar to the way Atlas from mythology held up the world.

ABBREVIATIONS AND ACRONYMS

Like many disciplines, the worlds of anatomy and physiology are full of shortcuts. After all, some of those descriptive terms grow quite large! You will often encounter modifications of full terms in the form of abbreviations or acronyms. You are already familiar with abbreviations. An **abbreviation** is a shortened form of a word or phrase. For example, the part of your digestive system that includes your stomach and intestines is referred to as the *gastrointestinal tract*, often abbreviated as the GI tract. Valves inside your heart regulate the blood flow between your heart's upper chambers, called atria, and its lower chambers, called ventricles. These valves are known as *atrioventricular valves*, which is often shortened to AV valves.

An **acronym** is technically a word formed from the first or key letters of each word in a multiple word name. It is pronounced as if it is a word. For example, AIDS is the acronym for **A**cquired **I**mmune **D**eficiency **S**yndrome, and is pronounced like the word "aids." Sometimes regular abbreviations are formed like an acronym by using only certain letters from multiple words. For example, HIV stands for human immunodeficiency virus, the cause of AIDS. But we pronounce the acronym, AIDS, as a word, while spelling out the abbreviation—H-I-V. Many sources lump abbreviations and acronyms together, and surely acronyms are a special type of abbreviation, but now you know the difference.

Abbreviations and acronyms can be problematic, though. SAD is an acronym for a condition called *seasonal affective disorder.* This disorder can be characterized as serious winter blues, apparently brought on by decreased daylight in winter months. The depression that accompanies it can be severe, making the acronym particularly suitable. However, a search of the medical abbreviations section of Medilexicon.com identified 21 terms that this abbreviation matches, and I found even more matches at other sites. They include Separation Anxiety Disorder, Social Anxiety Disorder, Small Airway Disease, Systemic Autoimmune Diseases, and Sporadic Alzheimer's Disease.

As you see, it is important to know the full name as well as the abbreviation, and to consider the context of how an abbreviation is used when determining its meaning. For example, an abbreviation used in medicine is AAA, referred to as a "Triple A," which stands for abdominal aortic aneurysm (**Figure 3.2**), a condition often seen in cadavers in an anatomy lab. This is a weakening and ballooning of the aorta, the body's largest artery, where it ends in the abdomen. If it ruptures, it can kill you quickly. But if I am en route to a meeting of AAA (the American Association of Anatomists) and my tire ruptures, Triple A or AAA—the American Automobile Association—will come and fix it!

Here are two last cautions about abbreviations: There may be more than one for the same thing, and some that sound quite similar may be different. For example, a *CAT* scan is an image created through a technology called computer-aided tomography, but these are now more commonly called *CT* scans. An *ECG* and an *EKG* are the same thing. Both abbreviations stand for electrocardiogram, a tracing of your heart's electrical activity—it doesn't matter if you use a "C" or a "K." But, if you change the middle letter to an "E," you are referring to an electroencephalogram, which is a tracing of your brain's electrical activity. Use abbreviations with caution, and don't worry much about them now—like eponyms, you will learn them as you go and it is easier to do so.

When using abbreviations and acronyms, always be sure that you know the full names and their meanings, and that you list the letters in the correct order. ■

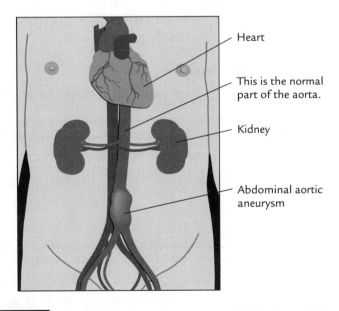

FIGURE 3.2 **An abdominal aortic aneurysm (Triple A or AAA).**

TIME TO TRY

Let's see how many of these abbreviations and acronyms you either know or can figure out. Match the following abbreviations to their full names.

1. _____ CBC

2. _____ ABC

3. _____ RBC

4. _____ ICE

a) Red blood cell

b) Ice, compression, and elevation— how you treat an injury to a joint to minimize the damage

c) Complete blood count—a lab test in which all the types of blood cells in a blood sample are counted

d) Airway, breathing, circulation—the order in which you assess a victim when administering first aid

These abbreviations probably gave you little trouble—you just had to match the letters in the name with the abbreviation. CBC means complete blood count, ABC is the assessment used for administering first aid, RBC is a red blood cell, and ICE is the acronym for ice, compression, and elevation.

✔ **QUICK CHECK**

1. What makes an acronym different from a regular abbreviation?

_____

2. SIDS stands for Sudden Infant Death Syndrome. Is this an abbreviation or an acronym? _____

Answers: 1. Unlike a regular abbreviation, an acronym is a word formed by key letters from the multiple words that it represents. 2. SIDS is an acronym, pronounced as it is spelled.

Putting Down Roots and **Building Descriptive Terms**

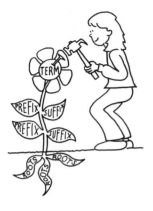

Most anatomy and physiology terms are descriptive and are built from two or more of three basic parts:

- A **prefix**, at the beginning of the word,

- A **root**, which is the main focus of the word, and

- A **suffix**, at the end of the word.

New words may be made simply by changing these parts (**Figure 3.3**). Some time ago, when learning English, you learned about prefixes and suffixes and how to use them to change one word to another, and we use them all the time. You will be amazed how many word roots, prefixes, and suffixes you already know.

A prefix is a short addition placed at the front of a word root. Consider the root *cycle*, which means circle or wheel. We can easily change the specific meaning of this root by adding prefixes as follows:

- uni*cycle*—has one wheel
- tri*cycle*—has three wheels
- bi*cycle*—has two wheels
- motor*cycle*—has a motor

You likely knew the prefixes we just used in our example. When you are reading anatomical terms, don't let the first glance worry you. Simply take a breath and look carefully at the word. Find the root, then examine any prefixes or suffixes that are used with it. Sometimes you may not know a particular word part at first, but you may figure it out just by thinking of words you *do* know that include that part. Let's try some.

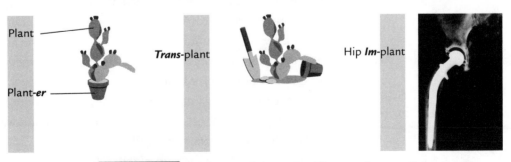

Plant

Plant-*er*

Trans-plant

Hip *Im*-plant

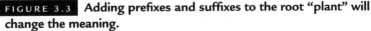

FIGURE 3.3 Adding prefixes and suffixes to the root "plant" will change the meaning.

TIME TO TRY

Here are some word parts and examples of words that use them. For each of the following, what do you think the word part means? Look at the examples and see if there is something they all share in common. This technique can help you determine the meaning of words with which you are unfamiliar—first look for the word parts and think of other terms that you do know that contain them.

Word part	Examples	Meaning
Root: cardi-	Cardiology, pericardium, cardiac arrest	_____ _____
Prefix: multi-	Multicellular, multitude, multilingual	_____ _____
Suffix: -ologist	Cardiologist, gynecologist, arthrologist	_____ _____

From this exercise, you probably can see that *cardi-* refers to the heart, *multi-* means many or multiple, and *-ologist* refers to one who is knowledgeable in or studies whatever the root says. *Cardiology* is the study of the heart, and one who studies the heart is called a *cardiologist.* Note how similar these terms are, yet their meanings are different. The *pericardium* is the sac that surrounds the heart (*peri-* means *around*). And *cardiac arrest* refers to when the heart abruptly stops functioning, such as from a heart attack.

Let's consider the name of your current course: anatomy and physiology. The prefix *ana* means up or apart, and the root *tome* means a cutting. *Anatomy* is a combination of the two. It is the science that focuses on the parts and organization of the body. But, when early folks tried to discover how we are built, all they saw was the outside. To discover how the body is organized, they did dissections—they *cut up* the bodies, *cutting away* outer structures to see deeper structures (**Figure 3.4**). See how descriptive the term really is?

Physiology comes from the root *physio*, which means nature or natural, and the suffix *–ology*, which means *the study of,* or *a branch of*

FIGURE 3.4   Early anatomists explored the body's structures by cutting away the outer structures to see inside.

knowledge. Physiology is the study of the branch of knowledge pertaining to nature. If you think of the natural sciences, which include chemistry and physics, this makes more sense—physiology is the study of how our body parts function, and that is largely determined by the laws of chemistry and physics.

Are you catching on to how easy this really is? Not only do descriptive terms convey a clear meaning, but they are easy to translate once you know the prefixes, suffixes, and roots. You already know several of these, and you will learn plenty more as you go. To help jump start your language acquisition, **Table 3.1** lists several common prefixes and their meanings, **Table 3.2** lists several common roots, and **Table 3.3** lists several common suffixes. Don't worry about learning them all now—they are there to help you start building your language skills. You should read through the terms and see which ones you already recognize.

Mastering word roots, prefixes, and suffixes is the most effective way to expand your A&P vocabulary because they can be recombined into countless new words. ■

TABLE 3.1 Some common prefixes.

Prefix	Meaning	Prefix	Meaning
a-, an-	without	infra-	below
ab-	away from	inter-	between
ad-	to or toward	intra-	within
alb-	white	iso-	equal
ambi-	both	later/o	side
ana-	up, back again, through-out, against	leuk/o	white
		macr/o	large
ante- or anter/o	before, in front of	mal-	bad
anti-	against	medi-	middle
bi-/bis-	twice, double	mega-	large
brady-	slow	melan/o	black
circum-	around	micro-	small
con-	with	mid-	middle
contra-	against	mono-	one
cyan/o	blue	multi-	many
di-	two, twice, double	neo-	new
dors/o	back	non-	not
dys-	hard, difficult, bad	para-	along side of, beside
ecto-	outside	per-	through, thorough, complete
erythr/o	red		
exo-	outside, outward	peri-	around, near
en-	in	poly-	many
endo-	within	poster/o	after, behind
epi-	on, upon	pre-	in front of, before
erythro-	red	pro-	before, in front of
extra-	outside, beyond, in addition to	pseudo-	false
		quad-	four
eu-	well, good, easy	retro-	backward
hemi-	half	semi-	half
heter/o	other	sub-	under, below
homo-	same	super-, supra-	above
hydro-	water	syn-	with
hyper-	over, above	tachy-	fast
hypo-	below, under	trans-	across
in-	in, into, on	tri-	three

TABLE 3.2	Some common word roots.

Root	Meaning	Root	Meaning
adeno	gland	gram	record
adipo	fat	graph	recording, writing
andr/o	male	gynec/o	female
angi/o	vessel	hem/o	blood
arter	artery	hepat/o	liver
arthr/o	joint	histo	tissue
aud or aur	hear	hydr/o	water
axilla	armpit	hyster/o, metr/o	uterus
brachi	arm	lapar/o	abdominal wall
bronch/o	lung air passageways	laryng/o	larynx, "voice box"
burs/o	bursa, "bag" (shock absorber between tendons and bones)	lumb/o	lower back
		mamm/o, mast/o	breast
		men/o	menstruation
cardi/o	heart	mening/o	membranes around the brain and spinal cord
caud/o	tail		
carp/o	wrist		
cephal/o	head	my/o, myos/o	muscle
cerebr/o	brain	nas/o	nose
cervic	neck	nephr/o, ren/o	kidney
chondr/o	cartilage	neur/o	nerve
clavi or cleido	clavicle	ocul/o	eye
colon/o	large intestine	oophor/o	ovary
cost/o	rib	op, opth	eye
cox/a	hip	orchid/o, test/o	testes (male gonad)
cubit	elbow	oste/o	bone
cyst/o	bladder, sac	oto	ear
cyt/o	cell	patho	disease
dent/o	teeth	pecto	chest
derm/o	skin	pes, ped, pod	foot
encephal/o	brain	phlebo or ven/o	veins
entero	intestine	plantar	sole of foot
gastr/o	stomach	pneumo/pulmo	lung
genesis	origin	procto	anus/rectum
gingivo	gums	pulmo/o	lung
glosso/linguo	tongue	ren	kidney
glute	buttocks	rhin/o	nose

▶

TABLE 3.2   Some common word roots, continued.

Root	Meaning	Root	Meaning
salping/o, salpinx	uterine tube	tom	cut
scope	examine closely	trache/o	trachea, "windpipe"
stasis	stay the same	ur/o, -uria	urine
stomato	mouth	vas/o	vessel, duct
talo	ankle	veno/phlebo	vein

TABLE 3.3   Some common suffixes.

Suffix	Meaning	Suffix	Meaning
-ac, -al	relating to	-ogen	precursor
-algia, -algesia	pain	-ologist	one who studies/ specializes in
-ase	enzyme		
-ate	do	-ology	the study of
-cide	kill	-oma	tumor (usually)
-cise	cut	-osis	full of
-cyte	cell	-ostomy	"mouth-cut"
-ectomy	removal of, cut out	-otomy	to cut into
-emia	condition of the blood	-pathy	disease of, suffering
-form	shaped like	-penia	lack
-genic	produced by	-plasty	surgical re-shaping
-gram	a recorded image	-plegia	paralysis
-graph/y	recording an image	-philia	affection for
-ia, -ic	relating to	-pnea	breathing
-ile, -illa, -illus	little version	-porosis	porous
-in	substance	-rrhage	excessive, abnormal flow
-ism	theory, characteristic of		
-itis	inflammation	-rrhea	discharge or flow
-ity	quality	-scopy, -scopic	to look, observe
-ium	thing	-sis	idea (makes a noun, typically abstract)
-ize	do		
-logy	study of, reasoning about	-stasis	to stop
		-stenosis	abnormal narrowing
-lysis	destruction, rupture	-tomy	cut
-megaly	large or enlarged	-ule	little version
-oid	resembling, image of	-um	thing (makes a noun)

TIME TO TRY

Match the following terms with their meanings.

1. _____ arthritis   a) Surgery through the abdominal wall

2. _____ otolaryngologist   b) Surgical removal of the appendix

3. _____ appendectomy   c) Difficult or painful menstruation

4. _____ dysmenorrhea   d) Inflammation of the joints

5. _____ laparotomy   e) Ear and throat doctor

How to Make a Jack-o'-Lantern: **Combining Word Roots**

When learning anatomical and medical terms, you should know some simple rules. You may have noticed in Table 3.2 that some roots end with "/o." That means the root can be used with or without an "o" added at the end. When the "o" is added, the modified root is referred to as the **combining form**. It is awkward to pronounce a word formed from a root that ends in a consonant letter and another root or suffix that starts with a consonant. For these roots, a vowel is inserted in the middle to make it easier to pronounce. This is most often an "o." For example, inside your body your abdominal cavity and your pelvic cavity are continuous, so they are often referred to as one cavity by combining the two roots: *abdomin* and *pelvic*. The resulting word would be *abdomin**np**elvic*. Instead, we insert an "o" between the roots to get *abdomin**o**pelvic*, which rolls off the tongue more easily. My students know this rule as the jack-o'-lantern construction.

TIME TO TRY

Combine the following word parts, then define the term.

1. gastr   +   -scopy = _____
    (stomach) (close examination)
    What does this term mean? _____

▶

2. rhin   +   -plasty = _____
    (nose) (surgical reshaping)
    What does this term mean? _____

3. hem   +   -rrhage = _____
    (blood) (excessive flow)
    What does this term mean? _____

4. encephal   +   -itis = _____
    (brain) (inflammation)
    What does this term mean? _____

5. cardi + myo   + pathy = _____
    (heart) (muscle) (disease)
    What does this term mean? _____

Did you get those? The first two combine to form *gastroscopy*, which means examining the stomach by inserting a viewing instrument (gastroscope) into it. The next pair form *rhinoplasty*, which is surgical reshaping of the nose—a nose job! The third pair combine to form *hemorrhage*, which means excessive blood, and refers to blood loss (bleeding). The fourth pair form *encephalitis*, which is inflammation of the brain. Note that this pair does not require the "o" for combining the two parts because the suffix begins with a vowel. Finally, you combined two roots and a suffix to form *cardiomyopathy*, which is disease of the muscles of the heart. Even though the root ends with a vowel, the "o" is still added—always be on the lookout for exceptions.

Here is another rule to remember: *Spelling counts!* In some cases, changing a single letter can make a big difference in the meaning of the word. For example, two terms that we use to discuss movement of body parts are *abduction* and *adduction*. Note that the only difference between the two is the second letter. However, abduction means to move a body part out to the side, or away from the midline, but adduction means moving towards the midline. Only that one letter distinguishes these two opposite terms. Here is another pair: *ilium* and *ileum*. The ilium is part of your hip bone, but the ileum is the last part of your small intestine.

Realize the importance of spelling—changing even one letter may change which structure you are naming. ■

TABLE 3.4 Basic rules for changing words in singular form to plural.

If the word ends in	Do this first	Then add	Examples
-a		Add -e	*vertebra* becomes *vertebrae*
-ax	Drop -ax	Add -aces	*thorax* becomes *thoraces*
-ex or -ix	Drop -ex or -ix	Add -ices	*cortex* becomes *cortices*
-ma		Add -ta	*dermatoma* becomes *dermatomata*
-is	Drop -is	Add -es	*anastomosis* becomes *anastomoses*
-nx	Change -x to -g	Add -es	*larynx* becomes *larynges*
-on	Drop -on	Add -ia	*ganglion* becomes *ganglia*
-us	Drop -us	Add -i	*nucleus* becomes *nuclei*
-um	Drop -um	Add -a	*ischium* becomes *ischia*
-y	Drop -y	Add -ies	*biopsy* becomes *biopsies*

Another maneuver that can be tricky with these terms is converting them from the singular form to the plural form. Some of the rules are the same as in English. If you look at one muscle and then another muscle, you are, indeed, examining two muscle**s**, and more than one bone are bone**s**. But there are some unique rules for pluralizing many terms that you will encounter. More than one vertebra are vertebra**e**, and the small bone at the end of each of your fingers is a phalanx, but all of these bones in your fingers together are called phalang**es**. **Table 3.4** shows the rules to guide you with pluralization.

Assume **The Anatomical Position**

Now that you understand how to read and form most of the terms used in anatomy and physiology, we can move on to some areas of specialized terminology that do not fit this pattern. The first of these to introduce is the **anatomical position.** This term refers to a specific body position that is used universally as a common reference point for the positions of body structures.

REALITY CHECK

If you are getting a bit tired of sitting, you might actually try the following positions. If you are feeling a little less active, just envision them.

1. You are lying on your couch or bed, face down.

What part of you is the top? _____

What is the bottom? _____

Which way is left or right? _____

Where is the front or the back? _____

2. You are stretched out in a recliner—upside down! Your feet are up over your head. Now what part of you is the top and what is the bottom? Next, sit in the chair the correct way. Do any of those positions change? _____

You likely see from this exercise that our perceptions of up and down, right and left, and front and back may change with changing body position. That is rather inconvenient if we are trying to describe the locations of structures within the body. To avoid such confusion, we always assume that the body is in the anatomical position, whether it is or not. I could just describe this position, but you will understand it better if you try it out, so on your feet please (this one is easy)!

TIME TO TRY

Stand straight and look directly forward. You are almost there already! Be sure the soles of your feet are planted firmly on the ground with your toes pointing forward. Let your arms hang straight down at your sides. In this natural position, your palms are probably facing your thighs. To move into the anatomical position, all you need to do is turn your palms so they are facing forward. Note this position (**Figure 3.5**)—you are now in the anatomical position! Remember it, because that is always the starting point when discussing anatomical relationships.

The anatomical position.

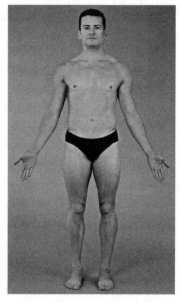

 All anatomical structures are named based on the anatomical position. ■

Not only does the anatomical position provide a common reference point, it also provides specific benefits. In this position, it is obvious that the limbs are parallel to the body and everything is straight. You do not see from the outside that there are two bones in the forearm—the *radius* and the *ulna*. When your palms are facing in toward your thighs as you usually hold them, these two bones are actually crossed, but in the anatomical position, they are parallel to each other. In this position, all the muscles, vessels, and nerves in the forearm are also running a straight course.

✔ **QUICK CHECK**

1. In what position are the hands when the body is in the anatomical position? _____

_____

2. Why is the anatomical position used? _____

_____

Answers: 1. They are facing forward. 2. It provides a common reference point for naming structures.

Anatomy ABC's: **Basic Anatomical Terminology**

Now that you know the anatomical position, we can introduce some terms that are used when discussing body structures. You will learn these, and plenty more, in class, but students sometimes struggle with these terms. You will have an advantage if you get started on them now.

RELATIVE POSITION TERMS

The following terms are often called relative position terms because they are used to compare the locations of structures. Many of these terms are paired, and that is the best way to learn them. When learning these, be sure to remember to always assume that the body is in the anatomical position.

Superior and **inferior**: *Superior* and *inferior* obviously mean above (toward the top) and below (toward the bottom), respectively. They have all your life, so you already know these terms. For example, your head is superior to your neck, and your thorax (chest) is inferior to your neck.

Cephalic and **caudal**: Look up these roots in Table 3.3 on page 107. *Cephalo-* means _____ and *caudo-* means _____, so these two terms mean toward the head or toward the tail. They are sometimes used in place of superior and inferior, which usually works fine. Just don't use caudal to refer to structures located below the tailbone.

Anterior and **posterior**: I am pretty sure you know these, too. They simply mean toward the front or toward the back. Your sternum is anterior to your lungs, for example, and your spine is posterior to them.

Ventral and **dorsal**: *Ventral* refers to the belly, and you already know dorsal—what is that thing on a shark that you see sticking up out of the water? _____ fin. Yes, that is the dorsal fin, and it is on the shark's back. *Dorsal* means toward the back. For example, your umbilicus (belly button) is ventral, and your spinal cord is dorsal. Most animals get around on all four legs, so for them these terms are synonymous with inferior and superior—that fin is on the top of a swimming shark, for example. But because we walk upright, our bellies are toward our fronts and our backs are, of course, toward our backs. So for us, *ventral* means the same as *anterior* and *dorsal* means the same as *posterior*.

Medial and **lateral**: Medial and lateral refer to positions along a side-to-side axis. Imagine an axis passing directly down through the center of your body all the way from the top of your head to the floor. *Medial* means towards the midline of the body, and *lateral* means to the side, away from the middle. A special term, **median**, refers to being exactly on the midline. Your heart is medial to your lungs, for example, and your arms are lateral to your trunk. Your trachea and esophagus are both in a median position.

Superficial and **deep:** You should already understand these as well. *Superficial* means closer to the surface, and *deep* means closer to the center of the body. In your head, for example, your skin is superficial and your brain is deep.

Proximal and **distal:** These terms are usually the most challenging for students, but they are actually rather simple. *Proximal* means closer to and *distal* means farther away. The question, though, becomes closer to or farther away from *what*? The reference point is some point of origin. This is easy for extremities—the point of reference is where the limb begins—where the limb arises from the trunk. So, in the upper extremity, the elbow is distal to the shoulder, but it is proximal to the wrist. Simply think of where the starting point is—here it is the attachment of the arm to the trunk. Then envision running your finger along the limb, starting where the limb starts. If you are comparing two locations, whichever you touch first is proximal to the other one. Thus, for the upper extremity the beginning is at the shoulder. Tracing out from there, the elbow is distal, or farther away from the starting point. You reach the elbow before you reach the wrist, so the elbow is *proximal* to the wrist—the elbow is closer to the beginning of the limb than the wrist is.

There is one other important pair of positional terms to remember, and you already use them routinely. Whether you are in class facing an anatomical model, standing over a cadaver, dissecting a cat, or treating a patient some time in the future, it is absolutely essential that you understand that *right* and *left* do not change. There is a tendency to think of right and left in terms of our own bodies, but if you look at someone who is facing you, your right side is facing their left side, and vice versa. Confirm this by standing face-to-face with someone while each of you raises your right arm. This may sound like a trivial thing, but every year horrendous medical mistakes are made because procedures are performed on the wrong side. Patients have even had the wrong limbs amputated. This is such a problem that many hospitals require doctors to mark with pens the surgical sites while the patient is still awake. Be sure you are always thinking of the *subject's* right and left, and not your own.

WHY SHOULD I CARE?

During your class, you will examine many structures whose names include relative position terms. When you take lab exams, you will be expected to know these names. For example, you have both a femoral artery and a deep femoral artery in your thigh. If you leave out the word "deep," you are incorrectly referring to an entirely different blood vessel. Similarly, the anterior and posterior tibial arteries supply different areas of the leg, as their names imply.

✔ QUICK CHECK

1. Your ankle is in what position compared to your toes? _____

2. Your nose is in what position on your face? _____

PLANES AND SECTIONS

We discussed previously that anatomy was originally revealed through dissections, which meant cutting into the body, removing parts, and cutting into the parts as well. Looking at structures from different angles gives us a better understanding of their true natures. But we then need words to help distinguish between these different views.

You can think of a **plane** as resembling a line that passes completely through the body in a particular direction, all the way from top to bottom or from side to side, and a **section** is a cut made along a certain plane. In class, you will discuss body parts that lie on certain planes within the body, and you will examine models and specimens that are cut in certain sections. Let's explore some of them now.

Answers: 1. proximal, 2. median.

FIGURE 3.6 **Major planes through the body: a)** sagittal, **b)** coronal, and **c)** transverse.

There are three major planes that pass through the body:

■ sagittal,

■ coronal, or frontal, and

■ transverse, or horizontal.

A **sagittal plane** passes vertically through the body from front to back, as shown in **Figure 3.6a.** You can think of it as dividing the body into right and left parts that may or may not be equal. If the plane passes directly through the midline of the body, it is called a **midsagittal plane** and it produces equal right and left parts. Recall, though, that a term that refers to the exact midline is *median*, so this same plane is also correctly called a **median sagittal plane,** or sometimes simply a **median plane.**

A cut made along this plane could be called a midsagittal section, a median sagittal section, or simply a median section. You will likely see this type of section in lab when you examine anatomical models such as the head or the pelvis that have been cut along this plane.

The second plane is the **coronal plane**, shown in **Figure 3.6b**. This plane also passes vertically through the body, but from side to side, so it gives you front and back parts. In fact, this plane is also known as a **frontal plane**. Your instructor may have a preference for which term you use, but I suggest to my students that they use frontal plane simply because it is easier to remember.

The third and final major plane is the **transverse plane** (**Figure 3.6c**). This is the only **horizontal plane**, and it is also referred to by that name, which is the one I prefer, again because it is easier to remember. This plane divides the body into top and bottom parts.

 PICTURE THIS

1. Imagine you are in Las Vegas at a magic show. Suddenly the magician pulls you out of the crowd and places you lengthwise into a large box. (If you can, please lie in that position now.) Much to your horror, he begins to saw you in half at the waist!

   Into what kind of parts are you separating? _____

   _____

   Which plane does this cut represent? Don't forget the anatomical position! Stand upright if it helps you get the answer. _____

   _____

2. Now, please stand in the anatomical position. Imagine there is a zipper running down the midline of your body from front to back, passing between your eyes, straight through your nose, and on down through your chin, neck, and so on. Now start pulling down the zipper.

   Into what kind of parts is your body separating? _____

   _____

   Which plane does this represent? _____

3. Again start from the anatomical position. You may have seen circus clowns act like they are running a string in one ear and out the other. Envision that, but change the string to a thin wire and start at the top of the head. Imagine doing that same side-to-side movement all the way down your body.

Into what kind of parts is your body separating? _____

_____

What kind of plane does this represent? _____

In the first part of this exercise, you should see that if the box is stood upright so that you are in the anatomical position, the magician's cut divides you into top and bottom sections and represents the transverse or horizontal plane. The zipper down your midline splits you into right and left parts and represents a sagittal plane—technically midsagittal, because it is on the midline. The final scenario splits you into front and back parts, representing a coronal, or frontal, plane.

Sometimes, though, we are looking not at the whole body, but rather at individual parts that have been cut open. Because they are cut, the particular view is referred to as a section. **Sagittal sections** are common through the head and trunk, but are less obvious in the extremities or in individual organs. For cuts through those, we more often use three different terms:

- longitudinal,
- cross, and
- oblique.

A **longitudinal section** is cut along the <u>long</u> axis (**Figure 3.7a**). A **cross section** is cut directly <u>across</u> the long axis and is also called a **transverse section** (**Figure 3.7b**). An **oblique section** is a section cut along any other angle (**Figure 3.7c**). I tell my students to associate the "o" in oblique with either the term "off" or "odd" because an oblique section is off of the main axes, making it the odd angle.

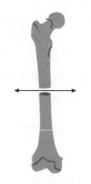

a) Longitudinal section   **b)** Horizontal or cross section   **c)** Oblique section

FIGURE 3.7   **Sections: a) longitudinal, b) transverse or cross, and c) oblique.**

TIME TO TRY

1. Obtain a long object through which you can cut, such as a carrot, a candy bar, or a drinking straw, and some scissors or a knife. (Be careful—sectioning yourself is not part of this exercise!)

2. Hold the object upright and envision the transverse plane. Now lay it down and cut along that plane about midway down the object. What kind of section did you just make? _____

3. Keeping the object in that position, cut one of the halves straight down the midline. What kind of section did you just make? _____

4. Take the remaining half and cut it at any angle different from the first two. What kind of section did you just make? _____

When you cut through the object in a transverse plane, you are making a cross-section. Cutting these halves down the midline makes longitudinal sections. Finally, cutting the object at any other angle produces oblique sections.

✔ **QUICK CHECK**

1. What are three specific names for a plane passing from front to back and directly through the midline of the body? _____

_____

2. A blood vessel is a tubelike structure. To remove a small section of a vessel, perhaps for a coronary bypass, you need to make two cuts. What kind of sections will you likely use? _____

In this chapter, I threw a lot of new terms at you. Most chapters in your textbook will do the same. At times, you may think that learning all the terms is impossible. Just when you think you know the name of a structure, you might learn that it has another name as well. Think about the red blood cell that carries oxygen in your blood. Do you know any other names for a red blood cell? You might—it is also called an **erythrocyte**, or it goes by the abbreviation RBC.

Multiple names are not uncommon in anatomy and physiology. In fact, my students often hear me say "Why name it once if you can name it two or three times and confuse people?" At times, the confusion may seem intentional, but realize that there are many ways to name things—formal names (erythrocyte) and informal names (red blood cell), for example. Mastering the language can be challenging, but you are now armed with important tools to help you. You will also find that many terms are defined in your textbook, either in the chapters or in the glossary. And you can always purchase a medical dictionary or refer to an online version—they are invaluable. If your schedule allows, you might consider also taking a medical terminology course while taking A&P—just don't overload yourself.

But remember the traveler who helped open our chapter. You now know to start your journey through each chapter in your textbook by learning the language. Knowing the terms will help you understand the material. Don't just memorize words—look up their meanings, look at

Answers: 1. Midsagittal, median sagittal, median. 2. Cross sections.

how they are built, make and use flashcards, and be able to use your new words. Don't get discouraged if you don't master all the terms in this chapter—those tables are long. Get in the habit of talking anatomy and physiology with your study group, friends, and family, for example. Learn what you can now and the rest will come as you go.

People often move to other countries and learn the language as they go by experience and practice. Students have been surviving in the foreign-sounding world of A&P for ages as well, and you have already had a crash course in the language. Enjoy your adventure!

Final Stretch!

Now that you have finished reading this chapter, it is time to stretch your brain a bit and check how much you learned. For online tests, tutorials, animations, activities, web links, and an ebook, visit the *Get Ready for A&P* companion website.

RUNNING WORDS

At the end of each chapter, be sure you have learned the language. Here are the terms introduced in this chapter with which you should be familiar. Write them in a notebook or enter them in your computer. Define them in your own words, then go back through the chapter to check your meaning, correcting as needed. Also try to list examples when appropriate.

Descriptive term	Inferior	Superficial	Coronal plane
Eponym	Cephalic	Deep	Frontal plane
Abbreviation	Caudal	Proximal	Transverse plane
Acronym	Anterior	Distal	Horizontal plane
Prefix	Posterior	Plane	Sagittal section
Root	Ventral	Section	Longitudinal section
Suffix	Dorsal	Sagittal plane	Cross section
Combining form	Medial	Midsagittal plane	Transverse section
Anatomical position	Lateral	Median sagittal plane	Oblique section
Superior	Median	Median plane	

WHAT DID YOU LEARN?

Try these exercises from memory first, then
go back and check your answers, looking
up any items that you want to review.
Answers to these questions are at the end
of the book.

**PART A: USING TABLES 3.1 THROUGH 3.3, MATCH
THE FOLLOWING TERMS WITH THEIR DESCRIPTIONS.**

1. _____ leukocyte

2. _____ endocarditis

3. _____ subclavian artery

4. _____ antecubital

5. _____ adipocyte

6. _____ mammography

7. _____ mammogram

8. _____ hysterectomy

9. _____ hepatomegaly

10. _____ quadriplegia

a) Paralysis of all four limbs

b) An image of the breast

c) An enlarged liver

d) Surgical removal of the uterus

e) A white blood cell

f) Inflammation of the lining inside the heart

g) A fat cell

h) An imaging technique used to assess the breasts

i) An artery located under the clavicle

j) The area in front of the elbow

PART B: ANSWER THESE QUESTIONS.

1. In what way is the anatomical position different from how you normally stand?

   _____
   _____
   _____

2. Which type of plane divides the body into right and left portions?

   _____

3. Which of the following is an eponym?

   a) CPR   b) SIDS
   c) cardiac sphincter d) McBurney's point

4. Which of the following is an acronym?

   a) CPR   b) SIDS
   c) cardiac sphincter d) McBurney's point

5. Which two planes pass vertically through the body?

   _____
   _____

6. Using Tables 3.1 through 3.3, construct a descriptive term for each of the following:

   A condition in which there are (too) many cells in the blood. _____

   A condition in which the liver is inflamed.

   _____

7. Using the tables, define each of the following terms:

   arterial stenosis _____
   _____

   chondrocyte _____
   _____

8. Provide the plural form of the following terms:

   pharynx _____

   mitochondrion (part of a cell)

   _____

   coxa (hip bone) _____

9. What is the difference between *medial* and *median*? _____
   _____
   _____

10. You decide to trim your fingernails into points by cutting at an angle from each side to the center. What kind of section are you making? _____
    _____

4 Body Basics

The Hip Bone's Connected to the . . .

When you complete this chapter, you should be able to:

■ Describe the Biological Hierarchy of Organization.

■ Explain the relationship between anatomy and physiology.

■ Discuss basic principles of biology and how they govern body functions.

■ Explain homeostasis and its importance to body functions.

■ Differentiate between tissues, organs, and organ systems.

■ List major organs in each organ system and understand each system's general functions.

Your Starting Point

Answer the following questions to assess your knowledge.

1. The study of tissues is called _____.

2. What is meant by "Form fits function?" _____

_____

3. Where does your body get the energy it uses for work? _____

_____

4. According to biological rules, what is the main reason for our existence?

_____

5. How many categories of tissues are there? _____

6. Which organ system secretes hormones? _____

Climbing the Ladder: **The Biological Hierarchy of Organization**

Now that you are armed with some new language skills, let's peek at what lies ahead in your studies of anatomy and physiology. You have a lot to learn along the way, but your journey is, literally, one of self-discovery—you are learning about yourself. That should make your adventures in A&P quite exciting! This chapter will introduce some basic concepts and principles to guide your path. Let's begin with how your course will likely be organized.

In science, we like to categorize. One classification system is known as the Biological Hierarchy of Organization. This hierarchy moves from the simplest level of structural organization up to the most complex. Several levels of this system are relevant to the disciplines of anatomy and physiology, and these are shown in **Figure 4.1**. Your textbook and

Answers: 1. Histology. 2. A part's structure reflects the job it does. 3. From food. 4. To survive and reproduce to continue the species. 5. Four. 6. Endocrine.

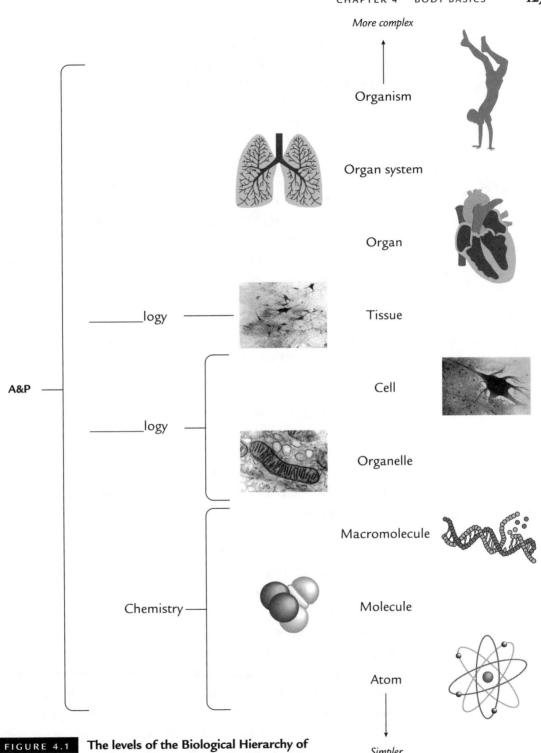

FIGURE 4.1   The levels of the Biological Hierarchy of Organization that are covered in anatomy and physiology.

your course will follow this pattern. These levels, from the simplest to the most complex, are the atom, molecule, macromolecule, organelle, cell, tissue, organ, organ system, and organism. (Technically, a macro-molecule is just a large and complex molecule, so some sources lump the two together as one level.)

TIME TO TRY

Examine Figure 4.1. Note that the first three levels are part of the science discipline called chemistry. The next two levels—organelle and cell—are covered by another science discipline. Look back at Tables 3.1 and 3.2 and determine the name of this science that studies cells and their structures. What is it? _____

Now, write that name in the appropriate line in Figure 4.1.

The next level is the tissue. Again, refer to Chapter 3 and determine the name of this branch of science.

What is it? _____.

Now add that label to Figure 4.1.

As you see from the illustration, the first three levels are **chemistry**, which we will review in Chapter 5. The next two levels are **cytology**, the study of cells, which we will explore in Chapter 6, and **histology,** the study of tissues. To study both cytology and histology requires magnification, typically through a microscope, so these disciplines are sometimes referred to as **microscopic anatomy**. The more complex levels—organ, organ system, and organism—will comprise the bulk of your anatomy and physiology course. Structures at these levels are visible without magnification, so their study is also called **gross anatomy**—not gross meaning *yucky*, but rather gross meaning *large.*

All matter is made of **chemical elements**. The smallest piece of an element is an **atom**, and atoms can unite to form **molecules**. To understand this, consider water, arguably the body's most important nutrient. One atom of the element oxygen combines with two atoms of hydrogen to form a molecule of water (H_2O). Water molecules are quite small, but some molecules, such as fat, DNA, starch, and proteins, are rather large, so they are referred to as **macromolecules**. Atoms, molecules, and macromolecules provide the nutrients and building materials our bodies need to stay alive and healthy, and they participate in chemical reactions that do all of the work performed in our bodies.

Macromolecules can unite to form complex structures called **organelles** that carry out functions inside cells. Organelles include cellular structures like the nucleus or a mitochondrion. **Cells** contain the combination of organelles necessary to sustain life, so cells are the first level of organization that we consider to be alive. In fact, cells are the basic units of all living organisms. **Tissues** are groups of cells organized to perform some common function. For example, muscle tissue contracts to provide movement. Tissues then organize into larger functional units called **organs**, such as your lungs, heart, and liver. Each organ performs at least one specific job. Multiple organs combine to form **organ systems,** each with some overall function. For example, the organs in your cardiovascular system are your heart and your blood vessels. The heart is the pump and the vessels are the pipes, so to speak, through which your blood travels to deliver nutrients and oxygen to your cells and to haul away their wastes. Collectively, your organ systems do the work needed to keep you—the **organism**—alive!

PICTURE THIS

You decide to build a house. You use lumber that came from trees, which were once living organisms. Explain how the first five levels of the Biological Hierarchy apply to the tree from which your lumber came. _____

_____

_____

_____

Trees are made of cells containing organelles that allow the tree to grow and form woody material, such as plant fiber, which is a macromolecule. Macromolecules are built from smaller molecules and they, in turn, are built from atoms.

Now let's use the idea of a house as an analogy for *you*, a living organism, to trace the structural progression from simple to complex. Various building materials (cells) are used to form the walls, floors, and ceilings (tissues). You add various systems such as plumbing to move the water in and your waste out, electricity to provide energy, and a heating and cooling system to regulate the air. The latter has a furnace, an air conditioner, and ducts. Each of these items has a unique job, so they are like organs. These "organs" form a complete system (the organ system). Once you get all these different "systems" in place, you have a fully functional house (the organism).

✔ **QUICK CHECK**

Fill in the missing levels from the Biological Hierarchy of Organization.
1. atom, molecule, _____, organelle
2. cell, _____, organ, _____, organism.

Some Things Never Change: **Basic Biological Rules**

I find most of anatomy and physiology to be beautifully simple and understandable, and I hope you will soon see what I mean. At first you may be tempted to just memorize the names and locations of anatomical structures, but you must move beyond memorization to truly understand the processes involved in the physiology. When studying the course material, always look for the logic and reasoning behind it. Your task will be easier if you understand some basic biological rules that govern the human body. These rules will help you understand why the body does what it does. There are, of course, exceptions to most rules, but these concepts should still help guide your thinking as you try to understand the "whys" behind your learning.

LIFE BEGINS AT THE CELL

Humans are made of trillions of cells, each of which is, or at least started as, a living unit. Many of our body functions occur within our individual cells. But our cells also have highly specialized functions and they

Answers: 1. macromolecule. 2. tissue; organ system.

communicate with each other, so many body processes also occur through the coordinated action of groups of cells. In fact, all of our organ systems are interrelated and work together to keep us alive. Always keep in mind that all body processes begin at the level of the cell.

FORM FITS FUNCTION

Recall that anatomy is the study of the body's *structure* and physiology is the study of the body's *functions*. Although these sound different, you should realize that they are closely related. You need to understand the parts and how they are put together to know how they work. Similarly, if you know what a body structure does, you can usually predict how it is organized to do its job. A common theme throughout biology is that "Form fits function." This means that all body parts have specific structures that allow them to perform their jobs most efficiently. A part's shape and organization reflect what it does, and similarly the job that is needed affects the structure that a part will have.

Let's consider the heart (**Figure 4.2**). The heart's job is to circulate your blood, and it does this by constantly pumping your blood along its way. To accomplish this, the heart has receiving chambers, the atria, and sending chambers, the ventricles. Some of the blood returning to the heart comes from the lungs carrying a rich supply of oxygen, and is ready to head out through your body to supply all your cells. The rest of the blood returning to your heart has just been to your cells and dropped off much of its oxygen for their use, so it needs to go to the lungs to get more. Fortunately, your heart is divided into right and left sides. This allows your blood to travel through two different paths—the right side sends the deoxygenated blood out to the lungs, and the left side sends the freshly oxygenated blood out to your body's cells.

Your blood circulates best if it is always moving in a single direction—forward. Doorways called valves are part of the heart's anatomy. One set allows blood to enter your ventricles and the other set allows it to leave these chambers. To fill the ventricles, the entrance doors are open but the exit doors are closed (**Figure 4.2b**). When the ventricles contract, the entrance doors close so the blood must move forward, pushing the exit doors open so it can leave (**Figure 4.2c**). Clearly the heart's anatomy is uniquely suited to its function. In other words, the heart's form fits its function.

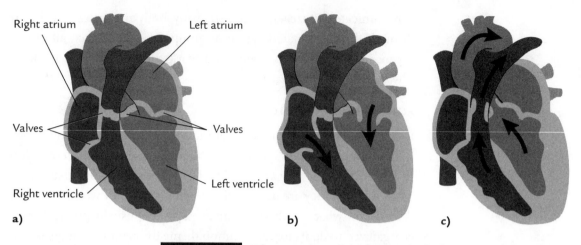

Right atrium

Left atrium

Valves

Valves

Right ventricle

Left ventricle

a)

b)

c)

FIGURE 4.2 **The anatomy of the heart. a)** The heart chambers and valves. **b)** As the ventricles fill, valves at the entrances into the ventricles are open but valves at the exits are closed. **c)** As the ventricles empty, valves at their entrances are closed but valves at their exits are open.

TIME TO TRY

Examine **Figure 4.3** to answer the following questions.

1. Figures 4.3 a) and b) show cross sections of thin and thick skin. Thin skin is found in most areas of your body, but the soles of your feet have much thicker skin. How do you think the thickness of the skin on the soles of your feet is related to the function of your feet? _____

_____

2. Let's try the opposite approach. You use your hands to do all kinds of work, some of it physically hard. The palms of your hands experience tremendous wear and tear. What type of skin do you think covers your palms? _____

3. Figures 4.3 c) and d) both show long bones, but the long bones in your hand are much shorter than the long bones of your forearm. How do the functions of these two areas explain this structural difference in their long bones? _____

_____

4. Figure 4.3 e) shows a top view of a thoracic vertebra from the vertebral (spinal) column. The vertebral foramen, shown here, is found in all vertebra. Why do you think that hole is there?

_____

FIGURE 4.3   **Form fits function. a)** Thin skin covers most of the body's exposed areas. **b)** Thick skin protects the soles of your feet and the palms of your hands. **c)** Long bones in the forearm. **d)** Long bones in the hand. **e)** A thoracic (upper back) vertebra from the spinal column.

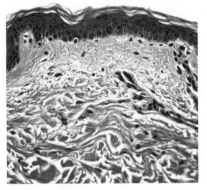

a)

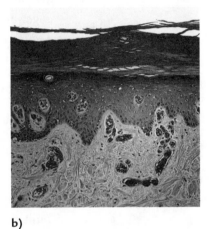

b)

Radius

Ulna

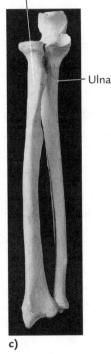

c)

Long bones

d)

Vertebral foramen

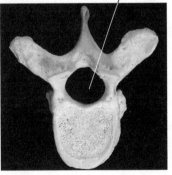

e)

From the previous exercise, you should realize that thick skin is located in areas that get the most use, pressure, and wear and tear. Thicker skin provides better protection for the underlying structures, and because your palms are used so extensively, they are covered by thick skin, like their counterparts in the lower extremity—the soles of your feet. Your forearm contains only two long bones, but there are 19 much shorter "long" bones in your hand. This allows the hand to move into many different shapes and to grip objects tightly, making it one of our most versatile tools. Vertebral foramina are found in all of your vertebrae, stacked atop each other to provide a hollow and protected passageway for your spinal cord and the nerves arising from it.

You already know many of your body's structures and what they do. As you learn the rest, always ask yourself the following questions:

- How is this structure built?

- What is its job and how is it performed?

- How do its structure and function fit together?

When you can answer those questions, you are truly learning anatomy and physiology. So when you examine a structure, the way it is built should give you clues about what it does. Similarly, if you know what a structure does, you should be better able to understand how and why it is built the way it is.

✔ **QUICK CHECK**

What is meant by "Form fits function?" _____

_____

_____

 A body part's structure reflects its function. ■

Answer: A part is organized in a manner that makes it most suited to the job it needs to perform. Its structure reflects it function.

LIFE REQUIRES ENERGY

All living organisms use energy. **Energy** is defined as the ability to do work, so it is required to do all of your body's work, or **metabolism**—virtually everything that is occurring in your body at any given time. Almost all energy on earth ultimately comes from the sun and is first harnessed by plants through photosynthesis. During this process, plants take in carbon dioxide and water from the environment and use solar energy to convert these molecules into chemical energy, which is stored as sugars. Plants also release oxygen as a waste product.

We cannot harness energy from the sun, so we eat the plants, or the animals that ate them, to get the chemical energy they created and stored (Thank you, plants!). Our cells then use that stored energy to perform our normal body processes. The process by which we convert stored energy into a usable form is called **cellular respiration**, and we get more energy out of our food if we use oxygen in this process. Much of this oxygen is released into our atmosphere by the plants as they perform photosynthesis. Then, as our cells harness the stored energy in our food molecules through cellular respiration, water and carbon dioxide are produced and released back to the environment so that they are again available to the plants that need these chemicals for photosynthesis. Body parts that are the most active (do the most work) have the highest energy demand, and we must continually supply them with energy in order for them to do their work.

REALITY CHECK

Based on your current knowledge of your organ systems, which two organ systems do you think help us obtain and use energy from food molecules? _____

_____

CONSERVATION

Without energy, we die, so our bodies are very careful about how we use our energy. Our cells conserve energy whenever possible. During the day, when you are awake and active, you use a large amount of energy to

Answer: The digestive and respiratory systems.

meet all your needs. But when you are sleeping, many of your organ systems don't have to work as hard, and you use less energy. If you find yourself in a potentially dangerous situation, your body uses more energy to prepare you to respond, perhaps by fleeing from the threat. But once you are out of harm's way, your energy use again drops. In almost any situation, the body uses the least amount of energy it can to do its tasks.

PICTURE THIS

Our energy comes from the foods we eat. You know that foods have Calories, but did you know that the **Calorie** is actually the basic measuring unit for energy? Think about our energy, or Calories, as being the money of life. To a certain extent, we can control our energy "income."

1. How? _____

2. If we do not take in enough energy "money," we cannot meet all of our energy demands, or, in effect, we cannot pay our bills. What happens in your home if you don't pay your power or your water bill? _____

3. What would happen to your organs if you did not give them enough energy? _____

4. If we earn more real money than we spend, what should we wisely do with the extra? _____

5. What do you think the body does with extra energy "money" it takes in? _____

Most extra energy, again meaning Calories, is stored in the body as fat, regardless of what we eat. This is the body's version of saving for a rainy day, and we gain weight. As with our bank accounts, to lose stored income (weight), we need to spend more than we are taking in. That may be, unfortunately, very easy to do with real money, but your body "bank" doesn't make it so easy to take those extra Calories out of storage because it is designed to conserve energy. It gladly lets you make deposits, but tries to keep you from making withdrawals. In fact, if you cut back dramatically on your caloric intake, your body becomes even more conservative and spends less energy, making it even harder to burn off those extra pounds.

JUST FOR FUN

Check out the nutrition label on your favorite food.

1. First, notice the serving size. Is that the amount you usually consume as a "serving"? _____

2. How many Calories are there in a single serving? _____

   Did you notice that food calories are listed as Calories with an uppercase "C"? This is a bit odd, but there is a difference between a calorie and a Calorie. *Huh?* The Calorie, with a capital C, is a food calorie, which is actually a kilocalorie. So, one food Calorie is really 1000 regular calories. Remember from the metric system that we use the most appropriate unit. For food it is the kilocalorie, but for some reason someone came up with the designation Calorie.

3. Knowing that 1 food Calorie = 1000 calories, how many calories are there in one serving of your favorite food? _____

   YIKES! One cup of skim milk, for example, has 90,000 calories (though I doubt that you picked that as your favorite food)! Don't panic though—those food Calories are the same ones you have always been consuming—all discussion of food uses the Calorie. But how odd it is that just changing the capitalization of the initial letter in the word makes a difference of 1000!

Obtaining all the materials we need to make and maintain our body parts requires considerable energy. To minimize this cost, our bodies recycle. Many molecules are broken down to atoms and then built back into other molecules, over and over again. Think about a child's building block set. Children can spend hours building things, taking them apart, then building new ones—a car is broken apart, then becomes a castle, which is broken apart and rebuilt into a robot. Like the child with its blocks, the body reuses both energy and materials.

I tell my students that the body is lazy—it prefers not to do anything it does not absolutely have to do. I further explain that, more accurately, the body is highly conservative—it doesn't make or maintain parts it doesn't need, and it spends as little energy as it can.

This theme of efficiency will help you understand many aspects of physiology. For example, the saying "Use it or lose it" is physiologically quite true. Tissues respond to repeated stress by building up and becoming stronger. Aerobic exercise stresses your heart, which responds by becoming stronger and more efficient. Strength training stresses your skeletal muscles, which respond by becoming bigger, stronger, and better at generating force. Weight-bearing exercise stresses the bones, so they thicken to become stronger. This is an important way for women to avoid the devastating consequences of osteoporosis (**Figure 4.4a**). But if you stop exercising and stressing these tissues, all the extra growth and development are quickly lost. Astronauts returning to Earth after being in a weightless state, for example, have lost measurable muscle and bone mass, leaving them slightly weakened upon their return. It takes energy and materials to maintain tissues, and the body's conservative nature usually prevents it from maintaining any that are not needed.

All living organisms require energy to do work, and the body is amazingly efficient and always tries to conserve energy and materials, avoiding wastefulness. ■

HOMEOSTASIS

To maintain the utmost efficiency, the body needs an optimal working environment. To achieve this the body has numerous mechanisms, most of which require energy, that work to maintain a relatively constant internal environment. This internal constancy is called **homeostasis**. You can think of it as maintaining the right balance of conditions in your body.

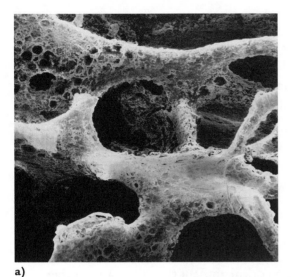

a)

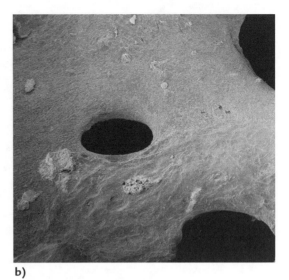

b)

FIGURE 4.4   **Bone tissue. a)** Bone that is not stressed during the first two decades of life is more likely to develop serious osteoporosis in later years. **b)** Healthy bone is dense and well developed and better able to resist age-related bone loss.

TIME TO TRY

Turn back to the tables in Chapter 3 to determine the literal meaning of the word *homeostasis*. _____

Homeostasis literally means "to stay the same." Most work done in your body is the result of chemical reactions. These reactions occur most efficiently if there is a relatively constant temperature, the right amount of water, and the right amount of chemicals, for example. We have an acceptable normal range of values for each of these. The normal range is also called the **set point.** We know that normal body temperature is around 98.6°F, so 98.6°F is the set point for body temperature. If our body cools too much, chemical reactions will occur more slowly or not at all. If our temperature gets too hot, chemical reactions speed up and some chemicals may be destroyed.

PICTURE THIS

Imagine you have some prized plants growing in your garden. Before leaving for two weeks during a brutal summer drought, you set a timer on your garden hose that will let it run at just the right flow for two hours each day while you are gone. You return home to find that your once-gorgeous flowers are now dried brown sticks! Some investigation reveals that although your water did run for two hours each day, the flow was reduced because some pesky varmint ate a hole in your special hose!

1. Why did your plants die? _____

_____

2. In the body, water is a critical nutrient that also carries other nutrients and waste products to and from your cells. What would happen to this transport if you had too little water in your body?

_____

3. How might that affect your cells? _____

_____

4. How would overwatering affect your plants? _____

_____

5. How might overwatering affect your cells? _____

_____

The human body is composed mostly of water. It is the main component of your cells and of your blood. Water balance is absolutely critical in the human body. If you have too little water, nutrients cannot be adequately transported to your cells and wastes can accumulate to toxic levels. Your cells will work less efficiently and may die. In plants, too much water dilutes the nutrients the plant cells need and causes them to swell and perhaps die. In the human body, too much water has the same effect, and when brain cells swell, death can occur. This condition is called water intoxication.

Let's consider another example. Think about how your home's furnace and air conditioner work. You set the desired temperature at, let's say, 70°F. If the air gets colder than that, your furnace turns on and heats the air until it is back to 70°, then the furnace shuts off. If the air temperature exceeds 70°, your air conditioner turns on and cools the air until it reaches 70° again, then it turns off. Notice that there are two possible problems—the temperature can go too low or it can climb too high. Both problems have a solution, but because the problems are opposites, so are the solutions—one raises the temperature but the other lowers it. Both solutions stop once the problem is corrected—they are self-limiting. This kind of control is called **negative feedback**, and it is the most common control mechanism used in physiology and the main way your body maintains homeostasis.

REALITY CHECK

Let's see what you know about your own body's furnace and air conditioning.

If you are overheated, how does your body respond to try to cool you down? _____

If you are too cold, how does your body respond to try to warm you up? _____

If you are too hot, you sweat more and the blood vessels in your skin dilate to bring the overheated blood to the surface of the skin where heat can be more easily lost. If you are too cold, you sweat less and blood stays away from the surface of the skin to keep your internal organs warmer. You shiver, which involves rapid contraction of your muscles. That activity generates heat.

Through negative feedback, body temperature stays close to its set point. Recall that a "normal" physiological value is a mathematical mean, so your own set point may be higher or lower than the "normal" value. Also, natural fluctuations occur throughout the day. Even with the set point varying a bit, negative feedback mechanisms usually control the body's activities to maintain homeostasis.

Negative feedback is our major, but not our only, control mechanism. **Positive feedback** works almost the opposite of negative feedback. With negative feedback, a body response is self-limiting, but positive feedback promotes further change away from the set point—the response increases. Think about parenting techniques. If a parent wants a child to stop doing something, negative feedback—perhaps a time out—usually stops the behavior. If a parent wants the child to continue a behavior, positive feedback rewards the behavior and increases the chance that it will continue.

Similarly, positive feedback in your body reinforces a change, causing even more change to occur. This usually takes the body away from the set point, often with tragic consequences. Let's again consider body temperature. As you overheat you sweat more, but if you sweat very long you can lose too much water. Water is a major component of your blood— too little water impedes the normal blood flow to your organs. To conserve water and keep your organs healthy, your body stops sweating if you lose too much water. Once that happens, positive feedback kicks in. Because you've suspended a major heat loss mechanism, your temperature rises. Increased heat speeds up the reactions within your cells, generating still more heat, so your temperature rises even more. Without intervention—actively cooling the body and replacing lost water—you may die from heat stroke.

Positive feedback isn't always bad. In childbirth, labor contractions trigger release of a hormone called oxytocin, which in turn causes the uterus to contract harder. As it contracts harder, even more oxytocin is released. The hormone and contractions reinforce each other with contractions coming faster and harder, hastening the baby's arrival—much to Mom's relief! When the baby begins breastfeeding, suckling releases the same hormone, which now increases milk flow. The more the baby nurses, the more milk it can drink.

WHY SHOULD I CARE?

Homeostasis and negative feedback are major recurring themes in anatomy and physiology. The normal state of the body provides optimal health. Any significant fluctuation away from the normal range can quickly impair normal function and perhaps become critical. These fluctuations form the basis of disease diagnosis. Negative feedback is the main control system by which our bodies correct these errors and restore health, so many of the processes we will discuss in the nervous and endocrine systems will deal directly with homeostatic control through negative feedback.

Balance, or homeostasis, is important in most aspects of physiology. Our bodies strive to stay within normal ranges for blood pressure, pulse, respiratory rate, oxygen and carbon dioxide levels, nutrient and waste levels . . . the list goes on and on. Much of what you will learn in physiology involves the mechanisms by which the body maintains homeostasis. Keep this concept in mind and you will be able to more easily predict how the body might respond in different circumstances and why the body does much of what it does. It's all about balance!

SURVIVAL

We have talked about many themes that govern how your body works, but the ultimate aspect is survival. I often tell my students that the body tries to keep us alive, often in spite of our own stupidity. Think about that—most of us can think of occasions when we did something that was not very healthy. Perhaps we overindulged in something we knew was not good for us. Maybe we tried some insane (and useless) fad diet or a dangerous daredevil stunt. Perhaps we went too long without sleep or got dangerously overheated while working outdoors in the summer. Although we *are* our bodies and minds, most of the brain functions below the conscious level to coordinate our bodies' activities, and to a great extent this is what keeps us alive even when we abuse ourselves.

TIME TO TRY

For each of the following situations, think about how your body might respond to try to survive.

1. You are stuck on an island with no food to eat. How will your body get the energy it needs to do work? _____

_____

2. You've been working outside on a hot day and you're sweating heavily. You are dehydrated. How will your body get its water balance back up? _____

_____

3. You have a badly infected finger. How does your body prevent you from dying from a bacterial infection? _____

_____

Our bodies are fairly resilient and rather forgiving. If you have no food, your body will start using the stored calories in your fat and other body materials to provide energy. Of course, it can only do that for so long before you starve, but most people can last for extended periods by drawing energy from their stores. When you are dehydrated, your brain triggers thirst so that you will replenish your water supply, and it also reduces sweating to conserve water. If you have a badly infected finger, your body has defense tactics, mostly conducted by your lymphatic (immune) system, that will fight off the infection.

There are limits, of course, to how much damage the body can withstand but, overall, the body has numerous control mechanisms and backup systems that can undo significant damage, self-inflicted or otherwise. Keep this in mind as you study physiology and you will better understand many of the reasons behind body functions. Homeostasis is maintained to ensure survival. We conserve energy and materials to ensure survival. Our parts perform optimally, making us phenomenally efficient machines with one goal—survival. It underlies almost everything our bodies do.

REPRODUCTION

Did you notice that I ended the last section by saying survival underlies *almost* everything the body does? It is only through reproduction that we pass on our genes and our species survives. Thus even that goal becomes one of survival—survival of more than just you and me. Many biologists maintain that the ultimate goal of our existence is reproduction. Indeed, the vast majority of differences between male and female form and function are tied to reproduction. And this is one area in which the body is not so conservative.

Females release at least one ovum, or egg, each month from the beginning of menstruation until menopause, but few of the ova are actually used. Likewise, each month the uterus builds up materials in preparation for a pregnancy that rarely occurs. Then, the newly formed materials are all flushed away as part of the monthly cycle. The levels of several hormones fluctuate in a complicated ballet each month just in case a pregnancy might happen, even for women who are celibate. Men make and release up to a half billion sperm per ejaculation, and if I know my physiology it only takes one to do the job! Now that does seem wasteful. The sexual act itself is also very energy-demanding.

In general, the more energy and materials the body puts into any job, the more important that work is. For this reason, it is clear why many scientists say that reproduction is our ultimate goal. But please don't feel compelled to rush out and do your biological duty—plenty of folks out there are doing this well enough to cover us! The goal is, after all, survival of the species, so some of us can sit out this dance without endangering the continuation of the human species. But while you are studying anatomy and physiology, keep this biological goal in mind.

Ultimately, your body functions are focused on your survival, and through reproduction the focus shifts from survival of you, the organism, to survival of the species. ■

✔ **QUICK CHECK**

1. What is homeostasis? _____

2. What is the most common way to regulate homeostasis?

_____

3. Biologically speaking, what is the goal of our existence?

_____

Answers: 1. Maintenance of a relatively constant internal environment. 2. Negative feedback. 3. Survival for the purpose of reproduction.

I Have to Learn All This? Does Anyone Have a **Tissue?**

Your course will probably begin with an overview of the characteristics of life, followed by a discussion of at least some of the basic rules we just covered. Then, you will move on to explore each of the levels of the Biological Hierarchy of Organization introduced at the beginning of this chapter. In the remainder of this chapter, we will superficially survey each of these levels. Our plan is to introduce these topics to arm you with some basic knowledge before you go into depth in class. In Chapters 5 and 6 of this book, we will go into a bit more detail on chemistry and cells. Skipping those levels for now takes us to our starting point for the remainder of this chapter: tissues.

Recall that life begins with cells and that groups of cells, in turn, form tissues with specific functions. Tissues must be examined microscopically, and there are four categories:

- epithelial tissue (epithelium),

- connective tissue,

- muscle tissue, and

- nervous tissue.

Epithelial tissues are like wrappings. They form all linings and coverings in the body. Name any part of the body that has a free surface and I will bet you it is covered by epithelium. The outer part of your skin, called the *epidermis*, is epithelial tissue. The outer surface of your heart, also called the *epicardium*, is epithelium. The linings of all of your hollow organs and your blood vessels are epithelia. But these linings often take on a special name—**endothelium**. *Epi-* is a prefix meaning upon, but *endo-* means inside, and since the epithelium that lines a structure is inside the structure, it is called *endo*thelium.

Connective tissue comes in many types (**Figure 4.5**) and one of its overall jobs, obviously, is to connect. One type of connective tissue, called loose connective tissue or areolar tissue, is basically the body's packing peanuts—it fills spaces between structures and is often specialized into a type of connective tissue called adipose, or fat. Dense fibrous connective tissue is very strong, and forms the tendons that anchor your muscles to

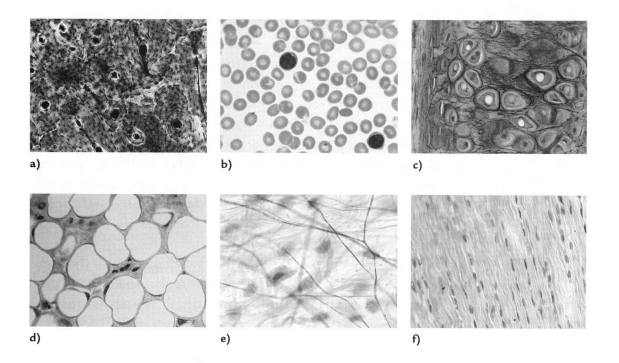

FIGURE 4.5   **Some types of connective tissue. a)** Bone; **b)** Blood;
c) Cartilage; **d)** Adipose; **e)** Areolar; **f)** Dense fibrous.

your bones and the ligaments that anchor your bones to each other. Cartilage is a shock absorber in many joints, where your bones meet. Bone itself is a connective tissue, as is blood, which is perhaps the body's ultimate connector. There are more types of connective tissue than of any other type, and this diversity allows for a wide array of functions.

TIME TO TRY

In what ways are bone and blood both connective tissues—what do they connect? _____

_____

Answer: Bone forms the skeleton upon which the whole body is built, and bones connect to each other and connect your extremities to your trunk, for example. Blood travels through the entire body and connects virtually every cell.

Muscle tissue comes in three types. The one with which you are most familiar is **skeletal muscle**. It provides body movement by attaching to two bones. When a muscle contracts, it moves the bones to which it is connected, thus moving your body parts or your whole body. This is the only type of muscle you can consciously control, so it is called voluntary. **Smooth muscle** is involuntary and is located in the walls of most of your internal organs and blood vessels. This muscle moves materials through these structures automatically, without you having the mental burden of trying to remember to do it. But the most specialized muscle is **cardiac muscle**, which is located exclusively in the heart, and is responsible for the contractions that continuously pump blood throughout your body.

Finally, **nervous tissue** is found in your brain, spinal cord, and nerves. This tissue sends signals to your organs and controls their actions. For example, it tells your muscles to contract, causes your glands to release their secretions, and allows you to think.

Can Anyone Here Play **The Organ**?

Now we are moving into areas with which you should be more familiar. Some of the organs you are likely aware of are the heart, lungs, brain, stomach, spleen, kidneys, and bladder. Each of these contains more than one type of tissue. Muscles, for example, are anchored to bones by tendons made of connective tissue and are covered by epithelial tissue.

Consider the heart (**Figure 4.6**). Both the outer covering and the innermost lining, as mentioned earlier, are epithelial tissues that provide protection as well as a smooth surface over which your blood can easily flow. The main bulk of the heart wall is cardiac muscle, which contracts. The heart also contains connective tissue, such as the valves that control the direction of the blood flow. Each of these tissues is needed for the heart to move your blood.

✔ **QUICK CHECK**

How can the words *muscle* or *bone* stand for both a tissue and an organ? _____

_____

_____

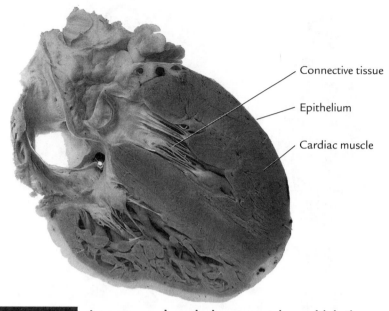

Connective tissue

Epithelium

Cardiac muscle

FIGURE 4.6   **An organ, such as the heart, contains multiple tissues.**

In Sickness and in Health: **The Organ Systems**

As you move through your anatomy class, you will do in-depth explorations of the individual organ systems. Our purpose here is to do a quick summary of each of them (**Figure 4.7**) to give you some background, to get you thinking about how these systems are integrated with each other, and to show ways in which the basic biological rules we previously discussed apply.

Answer: Muscle tissue refers to any of three types of tissue that can contract and produce movement. A skeletal muscle is an organ, and it is composed mostly of muscle tissue, but it also has connective tissue (the tendon) and epithelium that covers it. Bone is one type of connective tissue, but bones are organs that contain other tissues as well, specifically other connective tissues and epithelium.

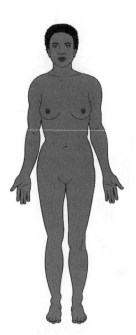

Integumentary system

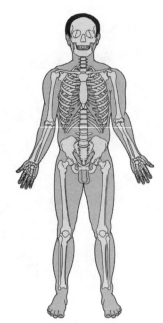

Skeletal system

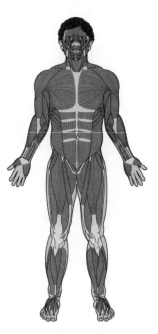

Muscular system

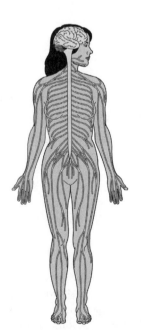

Nervous system

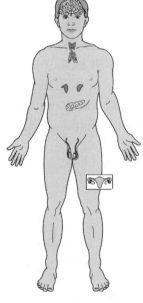

Endocrine system

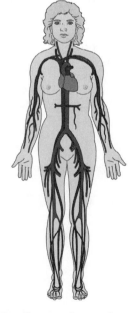

Cardiovascular system

FIGURE 4.7 The human organ systems.

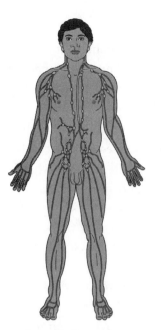

Lymphatic system

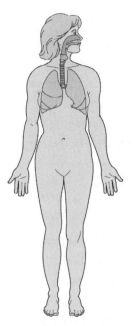

Respiratory system

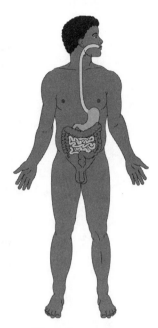

Digestive system

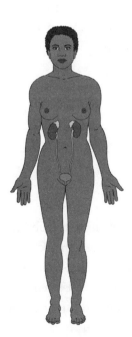

Urinary system

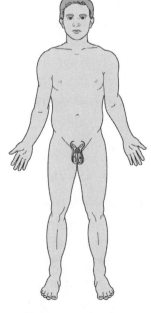

**Male reproductive
system**

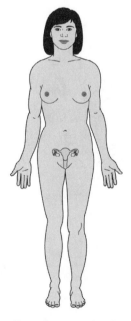

**Female reproductive
system**

THE INTEGUMENTARY SYSTEM

Your **integumentary system** includes your skin, hair, nails, and sweat glands. This system forms your body armor, providing a tough barrier that separates your inside world from the outside. This separation makes it easier to maintain homeostasis. The skin provides protection from potentially damaging substances such as bacteria or harsh chemicals. It helps regulate body temperature, as we discussed earlier. It is also loaded with sensory neurons that gather information about what is going on outside of you, alerting you if there is a potential threat.

THE SKELETAL SYSTEM

Your **skeletal system** is primarily composed of your bones, but it also includes the ligaments that hold them together, as well as cartilage and bone marrow. This system is the scaffold upon and around which your body is built. Your muscles attach to it to produce body movement. It also protects some of your most important organs, such as your brain, heart, and lungs, thus improving your odds of survival. Minerals, such as calcium, are stored in your bones. If the amount of calcium in your blood drops too low, calcium is released from the bones to maintain homeostasis. And your blood cells are made in your bone marrow, so this system is critical to your cardiovascular system.

REALITY CHECK

List any bones you know by name. _____

_____

THE MUSCULAR SYSTEM

Although there are three types of muscle *tissue*, when we talk about the **muscular system** we are usually referring to the skeletal muscles—the *organs*—and the tendons that anchor them to your bones. Your muscular system, obviously, can move your whole body and also parts of it. Your muscles also protect other structures, and they produce heat, such as when you shiver, that helps maintain your body temperature. Your muscles also consume much of your energy.

THE NERVOUS SYSTEM

Your **nervous system** is composed of your brain, your spinal cord, and the nerves that come off of them. This is one of your body's control systems. It continuously collects data in the form of sensory information about everything going on inside you and around you. It decides if any of that incoming information requires a response and, if so, directs that response, perhaps causing muscles to contract to move you away from potential harm, or maybe causing glands to secrete chemicals that digest your food. The nervous system also controls our special senses of taste, touch, smell, vision, hearing, and balance. This system is involved in almost every body function and has the major role of coordinating all of our organs and organ systems, keeping them functioning properly and working together to maintain homeostasis and keep us alive.

THE ENDOCRINE SYSTEM

Your **endocrine system** includes your pituitary, pineal, thyroid, parathyroid, and adrenal glands, your pancreas, and your gonads—ovaries or testes—as well as other organs. This system secretes hormones, which are chemical signals that control many aspects of your physiology. It is involved in normal body growth and development, and in maintaining homeostasis, for example, of water and minerals, glucose, and blood pressure, to name just a few. Hormones coordinate the functions of your various organs and regulate your body's metabolism, which is the rate at which you use energy. They also control your reproductive activities and cycles. Your endocrine system is intimately linked to your nervous system; in fact, the two systems together reign over most body functions and help you survive.

TIME TO TRY

For each of the hormones listed, write anything you know about what they do in your body.

1. insulin _____

_____

2. testosterone _____

_____

3. adrenaline/epinephrine _____

_____

4. A diuretic causes the body to lose water, usually by increasing urination. What do you think the hormone called *antidiuretic hormone* does? _____

_____

THE CARDIOVASCULAR SYSTEM

As mentioned earlier, the **cardiovascular system** includes your heart, your blood vessels, and the blood within them. This system circulates your blood, which carries nutrients, hormones, water, oxygen, and wastes, delivering the good stuff to the cells and carrying the bad stuff elsewhere in the body so it can be recycled or discarded. These functions are essential for homeostasis, growth and development, and reproduction. This system also helps regulate your body temperature and provides a common link between all cells. And we all know what happens to the organism if the heart quits.

Answers: 1. Insulin regulates blood glucose (sugar) by decreasing it, moving glucose into the cells to be stored for later use. 2. Testosterone is a sex hormone found in highest concentrations in males, causing many sex characteristics associated with maleness, and also contributing to the sex drive in both genders. 3. Adrenaline or epinephrine gives you that adrenaline rush—increased pulse, respiration, blood pressure—and makes you keenly alert and prepared for action in times of crisis. 4. Antidiuretic hormone causes the body to retain water—it has the opposite effect of a diuretic.

REALITY CHECK

List any blood vessels you know by name. _____

_____

THE LYMPHATIC SYSTEM

Your **lymphatic system** contains your spleen, thymus, tonsils, lymphatic vessels, and lymph nodes. This system has two major jobs. It is an alternative return route that picks up extra fluid and materials from your tissues and returns them to your blood, helping to maintain your blood volume and blood pressure. But the function of which you are most likely aware is your lymphatic system's role in keeping you healthy. This system is also referred to as your *immune system*, and it helps fight infection and disease.

THE RESPIRATORY SYSTEM

Your **respiratory system** includes your nose, pharynx (throat), larynx (voice box), trachea (windpipe), bronchial tree, and lungs. This system is designed to bring air into and out of the lungs so that the blood can drop off the carbon dioxide it picked up from the cells and pick up a fresh supply of oxygen. Recall our earlier discussion about cellular respiration. Blood from your lungs returns to your heart and shoots out through your body so that the oxygen it just gained can be used by your cells to harness energy from your food, allowing your cells to do their work. This cellular process generates carbon dioxide that is picked up by the blood and carried back to the heart to be routed back to the lungs to exchange gases once again—carbon dioxide out, oxygen in. From this, you can see that your respiratory system is directly tied to your cell's energy use. In addition, air that passes across your vocal cords causes them to vibrate, producing sound—perhaps so you can sing love songs to your sweetie with that goal of reproduction in mind!

THE DIGESTIVE SYSTEM

Your **digestive system** includes your mouth, salivary glands, pharynx (throat), esophagus, stomach, pancreas, liver, small intestine, and large intestine. This system brings food into your body and breaks it down to small molecules that enter your blood. These molecules then provide nutrients, building materials for growth and repair, and the chemical energy that your cells use to do your body's work. The digestive system also allows the body to get rid of wastes from the liver, as well as whatever you eat that is not absorbed into your body.

THE URINARY SYSTEM

The **urinary system** includes two kidneys, two ureters, the urinary bladder, and the urethra. This system filters your blood, getting rid of waste products and keeping what the body needs, thus maintaining homeostasis. It is crucial for maintaining the proper balance of water, minerals, glucose, blood volume and pressure, and pH in your body. Its role in ridding the body of wastes led to its alternative name—the excretory system.

✔ QUICK CHECK

The respiratory, digestive, and urinary systems all allow the body to get rid of wastes. How do all three systems do this? _____

_____

_____

THE REPRODUCTIVE SYSTEM

Finally, your **reproductive system** has different organs depending on your gender. Males have testes, epididymes, the vasa deferentia, seminal vesicles, prostate gland, penis, and scrotum. The male reproductive system produces sperm to fertilize the female's ovum, making Dad's contribution to the baby's genetic makeup.

The female reproductive system includes the ovaries, uterine tubes (oviducts), uterus, vagina, clitoris, labia, and mammary glands. A woman's system is designed to produce the ova that can be fertilized to

Answer: The respiratory system gets rid of the waste gas, carbon dioxide, when you exhale. Through defecation, the digestive system gets rid of wastes from the liver and the leftovers from eating that our bodies don't absorb. The urinary system cleanses our blood of wastes and excess materials through urination.

create a new life, with Mom contributing the other half of the baby's genetic makeup. If pregnancy occurs, her uterus becomes the baby's home during its initial nine months of development, and her mammary glands can feed the baby after it is born.

Together, all of your organ systems carry out all the functions needed to keep you—the organism—alive and, usually, healthy. These systems are all coordinated and interrelated, and illness or malfunction in one often leads to imbalance and malfunction in another. The human organism is, at once, a marvel of complexity and a showcase of simplicity, unsurpassed in its precision by any man-made machine. Take great pride in who you are, for you are a chemical and mechanical miracle!

Final Stretch!

Now that you have finished reading this chapter, it is time to stretch your brain a bit and check how much you learned. For online tests, tutorials, animations, activities, web links, and an ebook, visit the *Get Ready for A&P* companion website.

RUNNING WORDS

At the end of each chapter, be sure you have learned the language. Here are the terms introduced in this chapter with which you should be familiar. Write them in a notebook or enter them into your computer. Define them in your own words, then go back through the chapter to check your meaning, correcting as needed. Also try to list examples when appropriate.

Chemistry
Cytology
Histology
Microscopic anatomy
Gross anatomy
Chemical element
Atom
Molecule
Macromolecule

Organelle
Cell
Tissue
Organ
Organ system
Organism
Energy
Metabolism
Cellular respiration
Conservation
Calorie
Homeostasis
Set point
Negative feedback
Positive feedback
Epithelial tissue
Endothelium
Connective tissue

Muscle tissue
Skeletal muscle
Smooth muscle
Cardiac muscle
Nervous tissue
Integumentary system
Skeletal system
Muscular system
Nervous system
Endocrine system
Cardiovascular system
Lymphatic system
Respiratory system
Digestive system
Urinary system
Reproductive system

WHAT DID YOU LEARN?

Try these exercises from memory first, then go back and check your answers, looking up any items that you want to review. Answers to these questions are at the end of the book.

PART A: ANSWER THE FOLLOWING QUESTIONS.

1. List, in order from most complex to simplest, the levels of the Biological Hierarchy of Organization that are relevant to anatomy and physiology. _____

_____

_____

2. Examine your foot. It transmits your body weight to the ground and supports you in an upright position. How does your foot demonstrate that form fits function?

_____

_____

3. Where do we get the energy we use for body processes? _____

_____

4. How is energy related to your body weight?

_____

_____

5. What is homeostasis and why is it important? _____

_____

_____

_____

6. What are the four general types of tissues?

_____

_____

PART B: MATCH THE TERMS ON THE LEFT WITH ALL OF THEIR DESCRIPTIONS ON THE RIGHT.

1. _____ muscle

2. _____ lymphatic

3. _____ epithelium

4. _____ organelle

5. _____ bone

6. _____ heart

7. _____ cardiovascular

8. _____ nervous

a) A functional part of a cell

b) An organ

c) A tissue

d) An organ system

e) Covers or lines body parts

f) Skeletal system

g) Controls most body functions

h) Keeps you healthy

5 Chemistry

The Science of Stuff

When you complete this chapter, you should be able to:

◼ Explain the different states of matter.

◼ Describe atomic structure.

◼ Read and understand the Periodic Table of Elements.

◼ Explain ionic, covalent, and hydrogen bonding.

◼ Describe polar molecules and their unique characteristics.

◼ Discuss basic inorganic and organic molecules.

◼ Explain how we use chemicals for building and for energy.

Your Starting Point

Answer the following questions to assess your chemistry knowledge.

1. The most basic unit of a chemical substance is the _____

2. Matter is defined as anything that _____

_____

3. What are the three most common subatomic particles? _____

_____

4. Which subatomic particles interact during chemical reactions?

_____

5. What is the molecular formula for water? _____

6. What are three common types of chemical bonds? _____

_____

7. What happens in anabolic reactions? _____

_____

8. What is meant by *organic* molecule? _____

_____

9. Are proteins organic or inorganic? _____

10. What is ATP? _____

Yes, we are going to tackle some basic chemistry, but relax—it is really not that difficult. Why do you have to learn chemistry? Anatomy is the study of all of the parts and materials in the body—all of the "stuff" that you're made of. And chemistry is the science that covers all of that stuff. Physiology is the study of how the body works, and all of the work done in the body involves chemical reactions. Chemistry is very much a part of our everyday lives. Some of you may have previously taken a chemistry class, but others will be new to this discipline. In this chapter, we'll explore basic chemistry concepts to give you a head start in A&P class.

Recall from the last chapter our discussion of the Biological Hierarchy of Organization. It is organized from the simplest level of organization to the most complex. The first three levels are part of chemistry so, as you see, chemistry forms the very foundation of anatomy and physiology (**Figure 5.1**).

Learning chemistry may seem tough at times because the terminology can be challenging. To see what I mean, just read the ingredient list on almost any food product. *Do we really eat all that stuff?* But don't let the words interfere with your understanding—much of chemistry is quite simple, even though it may not seem so at first. Here's an example. What do you know about a chemical compound called *dihydrogen oxide*? You probably know more than you realize—that is the technical name for something we usually call *water*!

What's the **Matter?**

Let's start with something you already know about: **matter**. All the "stuff" of which you are made is matter, and matter is defined as anything that

- has mass (or weight), and

- takes up space.

The terms **mass** and **weight** are often used interchangeably, but there is a difference. Mass refers to the actual physical amount of a substance. Weight takes into account the force of gravity acting on that mass. Consider astronauts. Each has a certain mass—the actual

FIGURE 5.1 **The Biological Hierarchy of Organization for anatomy and physiology.** The simplest level of organization is the atom. The first three levels of this hierarchy are part of the discipline of chemistry: the science of matter. All of the other levels of organization are built upon this chemistry foundation.

More complex

Organism

Organ system

Organ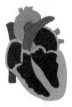

Tissue

Cell

Organelle

Chemistry

Macromolecule

Molecule

Atom

Simpler

amount of material his or her body contains. Each astronaut is weight-less during space travel when there is no gravity, but his or her individual mass does not change. For the sake of our discussion, matter can be defined using mass or weight, but mass is more precise. The second part of the definition of matter is that it takes up space. The space it occupies is called **volume**.

Matter typically exists in any of three physical states: solid, liquid, or gas. Once again let's consider water. What do we call the three states of water?

Solid: _____   Liquid: _____   Gas: _____

I hope you got those! Solid water is ice, liquid water is water (duh!), and its gas form is vapor, or steam. See—you already know chemistry! Now, how can you change solid water to its gas form? _____

_____

When you add heat, which is a type of energy, ice melts to become a liquid. With enough heat, the liquid eventually boils to become vapor. If you collect the vapor and cool it, it will condense back to liquid. If you cool it enough, it will become ice (solid). As you can see, the three forms of matter are interchangeable.

✔ **QUICK CHECK**

What are the three states of matter? _____

_____

It's **Element**-ary, My Dear Watson!

All matter is composed of **elements**, which are the most basic chemical substances. Over 110 elements are recognized, and around 90 of these occur naturally on Earth. Some elements you likely know are iron, copper, silver, gold, aluminum, carbon, oxygen, nitrogen, and hydrogen. Some exist in pure form, such as helium and neon, but most occur combined with other elements.

Answer: Solid, liquid, gas.

For the most part, living organisms require only about 20 elements. By weight, 95% of the human body is composed of just four of these:

- carbon,
- hydrogen,
- oxygen, and
- nitrogen.

Each chemical is represented by a symbol, typically the first one or two letters of the element's name. If more than one element name begins with the same letter, the most common of these elements usually gets the single-letter symbol. Hydrogen, for example, is represented by H, and the less common helium is prepresented by He. The symbols for some of the elements, such as the four listed above, are quite logical. Others are less obvious. For example, the symbol for silver is *Ag*, but that is because it comes from the Latin word *argentums*, meaning *silver*. **Table 5.1** lists the names and symbols of some of the elements that are most important for life.

TABLE 5.1 **Some of the important elements in living organisms.**

Element Name	Chemical Symbol	Element Name	Chemical Symbol
Hydrogen	H	Phosphorus	P
Carbon	C	Sulfur	S
Nitrogen	N	Chlorine	Cl
Oxygen	O	Potassium	K
Sodium	Na	Calcium	Ca
Magnesium	Mg	Iron	Fe

TIME TO TRY

Several elements have names that begin with the letter C, so most of them use a two-letter chemical symbol. Try to match each of the following chemical names with their symbols. (*Hint: Recall that one of these is very common and is a major component of all living organisms, including the human body.*)

Your Choices	Names	Symbols
_____	Calcium	Cu
_____	Chromium	C
_____	Cobalt	Ca
_____	Copper (*Latin = cuprum*)	Co
_____	Carbon	Cr

Chemical Carpentry: **Atomic Structure**

All chemical elements are composed of tiny particles called **atoms**. An atom is the smallest complete unit of an element—one atom of carbon, for example, is the smallest unit, or piece, of carbon that can exist. Two or more atoms can combine together to form larger structures called **molecules**. And simple molecules can join together to form more complex chemical structures called **macromolecules**. These include things like proteins, carbohydrates, DNA, and fats—many of the substances we associate with living organisms. But they all begin the same way—with atoms.

Atoms vary in size, weight, and how they interact with other atoms, but they all share some common characteristics. All are made of smaller units called **subatomic particles** that are arranged in a very precise manner. Although many subatomic particles are now recognized, the main ones of interest to us are **protons**, **neutrons**, and **electrons**.

THE NUCLEUS

The **nucleus** of an atom is not a structure. Instead, think of the nucleus as the area in the middle of an atom where some of the subatomic particles hang out. This can be confusing, because the nucleus of an atom is often referred to as if it is a structure. You should merely think of it as the atom's central region.

Answers: Calcium = Ca; Chromium = Cr; Cobalt = Co; Copper = Cu; Carbon = C.

An atom's nucleus is where we find two types of relatively large sub-atomic particles called **protons** and **neutrons**. Protons and neutrons have a similar size and about the same mass. Protons are positively charged particles and may be designated as p^+. Neutrons carry no electrical charge and they are, as their name suggests, neutral. Neutrons may be designated by n^0, indicating they lack any electrical charge, or simply by **n**. All of the protons and neutrons in an atom are located in the nucleus.

ELECTRONS

Orbiting around the nucleus are the other major subatomic particles—the **electrons**—that are in constant motion. Electrons are very small and have almost no weight. They also carry a negative charge, and are often designated as e^-. Because the protons are inside the nucleus, the nucleus always has a positive charge. However, the number of negatively charged electrons orbiting the nucleus always equals the number of protons at the nucleus. Thus the negative charges of the electrons exactly balance the positive charges of the protons. That means that any atom is, over-all, neutral.

An atom is electrically neutral. The number of e^- = the number of p^+. ■

Electrons are never in the nucleus. Let's start with a simple image to get our bearings. Think of the rings of the planet Saturn. The rings never touch the planet itself. These rings can represent the paths of the electrons around the nucleus. Unlike Saturn's rings, however, the electrons do not travel in a nice, even, straight line along a single plane. Rather, they buzz about quite rapidly in multiple paths called **orbitals** that have different shapes and different orientations. For this reason it is more accurate to envision a cloud of electrons that constantly circle the entire nucleus. A picture of an electron cloud does not show actual electrons. Instead, it is more like a map that shows the probability of where the electrons are at any moment.

PICTURE THIS

You arrive home late one evening, well after the sun has set. Your front light is on so you can see to put your key in the lock. You glance overhead and see a large cloud of insects swarming around the light. This is the basic image you should have of the electrons orbiting the nucleus of an atom (**Figure 5.2**). They are constantly in motion around the nucleus, but their individual paths vary.

Now that your skin is crawling from thinking about insects swarming overhead, let's get more specific. The movement of electrons around the nucleus is not as random as the movement of the insects around the light. It is actually a bit complicated, but for our purposes a simplified version will do. Electrons circle around the nucleus at different energy levels, each of which is called a **shell**, and these each have their own paths around the nucleus. Each shell has its own range of distance from the nucleus and its own path shape, so you can think of each shell as being a specific part of the electron cloud. The first shell is closest to the nucleus. Electrons try to stay as close to the nucleus as they can, but the first shell can only accommodate two electrons. If an atom has more than two electrons, it must have more than one shell. Each additional shell is located a bit further from the nucleus. Electrons always fill the shells from the nucleus outward.

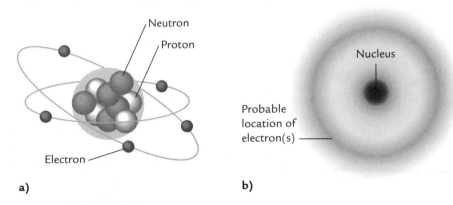

FIGURE 5.2 **Electron orbitals. a)** Electrons orbiting the nucleus of an atom of boron. **b)** An electron cloud shows the probable location of the electron(s) at any given time.

✔ **QUICK CHECK**

What are the three subatomic particles, and where is each located in an atom? _____

_____

_____

ATOMIC NUMBER

Each element has its own **atomic number**. It is, by definition, the number of protons in an individual atom of that element. Each element has a specific number of protons, and all atoms of that element have the same number. For example, hydrogen has an atomic number of 1. If you gave hydrogen another proton, it would no longer be hydrogen—it would now become a different element—helium, with an atomic number of 2. So, in order for atoms to be of the same element, they must all have the same number of protons.

Atomic number = number of protons in an atom. ■

Now, recall that an atom is electrically neutral overall. This means that the number of positive charges from protons must be counterbalanced with an equal number of negative charges from electrons. In other words, the number of protons in an atom always equals the number of electrons. Because of this, if you know an atom's atomic number, you know not only how many protons it has, but also how many electrons it has—they are the same numbers! Let's try this.

TIME TO TRY

Nitrogen's atomic number is 7.

1. How many protons does an atom of nitrogen have? _____

2. How many electrons does an atom of nitrogen have? _____

Answer: Protons and neutrons are always in the nucleus; electrons are always orbiting around the nucleus.

If nitrogen's atomic number is 7, that tells you it has 7 protons. You know the number of protons must equal the number of electrons, so an atom of nitrogen will also have 7 electrons. The seven positive charges from the protons are balanced by the seven negative charges from the electrons, so the atom is electrically neutral.

The number of protons = the number of electrons in an atom. ■

ATOMIC WEIGHT OR MASS

An element's **atomic mass** refers to the total mass of a single atom of that element. Recall our earlier discussion, though, about mass and weight: the terms **atomic mass** and **atomic weight** are used interchangeably. Because you are likely more familiar with weight than mass, we will stick with *atomic weight*. You need to recognize, though, that

1. either of these terms can be used

2. atomic mass is the more precise term, and

3. they have almost the same basic meaning.

For atomic weight we are really talking about how much a single atom of a particular element weighs. Electrons are so tiny that they weigh almost nothing. Almost all of the atomic weight, then, comes from the combined weights of the larger protons and neutrons. But what do *they* weigh? Obviously atoms are too tiny to be weighed in pounds or ounces, so imagine trying to weigh a subatomic particle! Conveniently, scientists developed a unit of measurement called the **atomic mass unit (u)**, and 1 u is, to simplify it, about the mass or weight of one proton. Neutrons are almost the same size, so we assume that they have the same weight as a proton. This becomes amazingly simple! To determine an atom's weight, all you do is add its total number of protons and neutrons together because each of them weighs 1 u, and you can simply ignore those tiny little electrons!

An element's atomic weight = number of protons + number of neutrons in an atom of that element. ■

For all atoms of an element, the number of protons is constant. But the number of neutrons can vary, so atoms of the same element can have different atomic weights. For example, carbon (atomic number 6) has 6 protons, and it usually has 6 neutrons, so its atomic weight is usually 12. But some atoms of carbon have 7 neutrons, giving an atomic weight of 13, and some have 8 neutrons, giving an atomic weight of 14. Atoms of the same element that have different atomic weights are called **isotopes**. The latter example is called "carbon 14." Isotopes are usually designated by the element's name and atomic number: C-14 or ^{14}C, for example.

WHY SHOULD I CARE?

Some isotopes are radioactive, meaning they emit certain types of energy. For this reason, some radioactive isotopes are used in medicine. The energy they emit can often be seen with special equipment. For example, the thyroid gland uses iodine to make certain hormones. If a patient might have a thyroid problem, a radioactive isotope of iodine (^{131}I) can be injected into the blood, then the clinician can use an imaging technique to monitor how well the thyroid is working. In another use, cobalt (^{60}Co) can be injected into an area where there is cancer to irradiate the tumor cells.

Is That An Eye Chart or a **Periodic Table of Elements?**

Look at **Figure 5.3.** YIKES! It may look a bit scary at first, but that's only because you don't know how to read it. This is the **Periodic Table of Elements.** All chemical elements that are currently known are listed in this table, and more are added as they are discovered. The table is arranged in a specific manner that is quite useful, so we will take some time to explore it.

Look at **Figure 5.4a.** From the Periodic Table, we know that hydrogen's atomic number is 1, meaning it has 1 proton, which is shown in the center of the atom at the nucleus. This means it also has 1 electron, which is shown orbiting the proton, in the first shell. It rarely has any neutrons. Helium's atomic number is 2. The helium atom in **Figure 5.4b** has 2 protons and also 2 neutrons (so what is its atomic weight? _____). Helium also has 2 electrons, as shown. Next, look at lithium

1A																	8A
1 **H** 1.008	2A											3A	4A	5A	6A	7A	2 **He** 4.003
3 **Li** 6.941	4 **Be** 9.012											5 **B** 10.81	6 **C** 12.01	7 **N** 14.01	8 **O** 16.00	9 **F** 19.00	10 **Ne** 20.18
11 **Na** 22.99	12 **Mg** 24.31	3B	4B	5B	6B	7B	⌐—8B—⌐			1B	2B	13 **Al** 26.98	14 **Si** 28.09	15 **P** 30.97	16 **S** 32.07	17 **Cl** 35.45	18 **Ar** 39.95
19 **K** 39.10	20 **Ca** 40.08	21 **Sc** 44.96	22 **Ti** 47.87	23 **V** 50.94	24 **Cr** 52.00	25 **Mn** 54.94	26 **Fe** 55.85	27 **Co** 58.93	28 **Ni** 58.69	29 **Cu** 63.55	30 **Zn** 65.41	31 **Ga** 69.72	32 **Ge** 72.64	33 **As** 74.92	34 **Se** 78.96	35 **Br** 79.90	36 **Kr** 83.80
37 **Rb** 85.47	38 **Sr** 87.62	39 **Y** 88.91	40 **Zr** 91.22	41 **Nb** 92.91	42 **Mo** 95.94	43 **Tc** (98)	44 **Ru** 101.1	45 **Rh** 102.9	46 **Pd** 106.4	47 **Ag** 107.9	48 **Cd** 112.4	49 **In** 114.8	50 **Sn** 118.7	51 **Sb** 121.8	52 **Te** 127.6	53 **I** 126.9	54 **Xe** 131.3
55 **Cs** 132.9	56 **Ba** 137.3	57* **La** 138.9	72 **Hf** 178.5	73 **Ta** 180.9	74 **W** 183.8	75 **Re** 186.2	76 **Os** 190.2	77 **Ir** 192.2	78 **Pt** 195.1	79 **Au** 197.0	80 **Hg** 200.6	81 **Tl** 204.4	82 **Pb** 207.2	83 **Bi** 209.0	84 **Po** (209)	85 **At** (210)	86 **Rn** (222)
87 **Fr** (223)	88 **Ra** (226)	89† **Ac** (227)	104 **Rf** (261)	105 **Db** (262)	106 **Sg** (266)	107 **Bh** (264)	108 **Hs** (269)	109 **Mt** (268)	110 **Ds** (271)	111 **Rg** (272)	112 — (285)	113 — (284)	114 — (289)	115 — (288)			

58 **Ce** 140.1	59 **Pr** 140.9	60 **Nd** 144.2	61 **Pm** (145)	62 **Sm** 150.4	63 **Eu** 152.0	64 **Gd** 157.3	65 **Tb** 158.9	66 **Dy** 162.5	67 **Ho** 164.9	68 **Er** 167.3	69 **Tm** 168.9	70 **Yb** 173.0	71 **Lu** 175.0
90 **Th** 232.0	91 **Pa** 231.0	92 **U** 238.0	93 **Np** (237)	94 **Pu** (244)	95 **Am** (243)	96 **Cm** (247)	97 **Bk** (247)	98 **Cf** (251)	99 **Es** 252	100 **Fm** 257	101 **Md** 258	102 **No** 259	103 **Lr** 260

FIGURE 5.3   **The Periodic Table of Elements.**

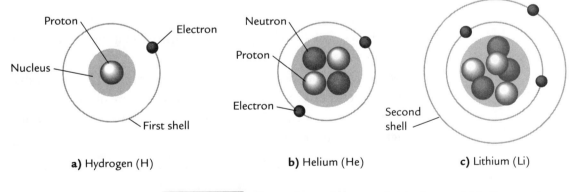

a) Hydrogen (H)   **b)** Helium (He)   **c)** Lithium (Li)

FIGURE 5.4   **Illustrations of atoms of a)** hydrogen, **b)** helium, and **c)** lithium show the placement of the electrons around the nucleus. The first shell fills first and can hold only 2 electrons.

(**Figure 5.4c**), which has atomic number 3. You see it has 3 protons and 3 neutrons in its nucleus, and 3 electrons in orbit. The first shell can only hold two electrons, so a second shell is added to hold the third electron.

TIME TO TRY

In the illustrations below, assume that the gray sphere in the middle represents the nucleus and all of the neutrons and protons in it. Add the shells and electrons for each of the elements below. Boron is done as an example. Don't worry about the positions of the electrons; just draw the right number in the right shell.

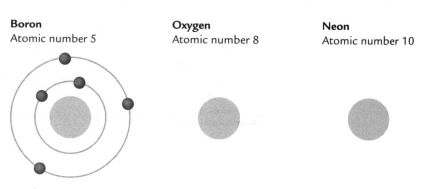

Boron
Atomic number 5

Oxygen
Atomic number 8

Neon
Atomic number 10

Needs 5 e⁻ total.
2 e⁻ are in the inner shell,
the other 3 in the outer shell.

In your drawings, oxygen should have 2 electrons in its inner shell and 6 in its outer shell. Neon should have 2 electrons in its inner shell and 8 in its outer shell, giving it a full outer shell.

Now look at **Figure 5.5**. This is the square from the Periodic Table that represents carbon. You can see that each square of the Table tells you an element's chemical symbol, its atomic number, and its atomic weight. What do those three items tell you? _____

_____

_____

Remember, if you know an element's atomic number, you know how many protons are in its atoms. Once you know that, you also know how many electrons it has, because the number of protons and electrons is the same. But look at the atomic weight. Carbon's atomic weight is 12.01.

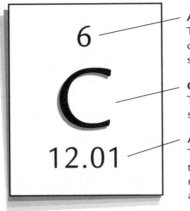

Atomic Number
This is the number of protons in one atom of this element. The atom will have the same number of electrons.

Chemical Symbol
This one- or two- letter abbreviation specifies which element this is.

Atomic Weight (Mass)
This is the weight (or mass) of one atom of this element. It is determined by adding the number of protons and neutrons together and is usually listed as an average value.

FIGURE 5.5   Information contained in the Periodic Table.

Atomic weight equals the number of protons plus the number of neutrons. How can you have 0.01 (1/100th) of a proton or a neutron? You can't. Remember isotopes? The atomic weights shown in the Periodic Table are averages of samples that contain isotopes. You can round up or down to a whole number to figure out the typical number of neutrons.

How would you do that? If carbon has an atomic number of 6 and an atomic weight of 12.01, how many neutrons does it typically have?

_____

The correct answer is 6, because the atomic weight is closest to 12, of which 6 are protons. The rest are neutrons.

TIME TO TRY

Use the Periodic Table (Figure 5.3) to answer the following questions.

1. How many protons are there in an atom of calcium (Ca)?

_____

2. How many electrons are there? _____

3. How many neutrons does a typical atom of calcium have?

_____

4. How many neutrons does a typical atom of phosphorus (P) have? _____

You should see that calcium's atomic number is 20, so it has 20 protons and thus also 20 electrons. Its average atomic weight is 40.08, which rounds down to 40. Of that, 20 are protons, which leaves 20 neutrons. Phosphorus has an atomic number of 15, meaning it has 15 protons. Its average atomic weight is 30.97, closer to 31. So,

$$31 - 15 \text{ protons} = 16 \text{ neutrons.}$$

See—this is just simple math!

Now look again at the Periodic Table. Starting at the top, read it across, left to right, row by row. How is it organized? _____

_____

_____

Next, look at the atomic numbers for the first four elements in Column 1— 1 (H), 3 (Li), 11 (Na), and 19 (K). Remember that in any atom the first shell holds only 2 electrons, and each of the next two shells can initially hold up to 8 more. Fill in the missing information:

Element: H   Li   Na   K
Electrons in its outer shell: _____   _____   _____   _____

You should see that all elements in Column 1 have a lone electron in their outer shell. Hydrogen has only 1 electron. Lithium has 2 electrons in the first shell and 1 in the outer shell. Sodium has 2 in the first shell, 8 in the second, and 1 in the outer shell. Potassium has 2, 8, 8, and 1. If you do the same for the second column, you'll find that each element has 2 electrons in its outer shell. And you'll find that all the elements in the last column have 8 electrons in their outermost shells, except for helium, which has only 2 electrons.

The Periodic Table is organized by atomic number. The atomic number—the number of protons—increases from the left to the right, and from the top to the bottom. The Table is also organized into rows, called **periods**, and columns, called **groups**. Each row, or period, represents a shell of electrons. The first row has one shell, the second row has two shells, and so on. Each column, or group, represents how many electrons are in the outermost shell. This organizational approach is very simple to use for the first three rows of the Periodic Table, but becomes more complicated below that. Those complexities, however, are beyond the scope of our current discussion.

✔ **QUICK CHECK**

What information do you know from the period in which an element is found in the Periodic Table? _____

_____

What information do you know from the group in which an element is found in the Periodic Table? _____

_____

Bumper Cars and **Chemical Interactions**

Have you ever tried bumper cars? If not, you really should—it's a great way to release tension. Atoms interact with each other rather like bumper cars do (**Figure 5.6**). The first part of a bumper car that makes contact with another car is the outer rubber bumper. When two atoms come together, the first parts to make contact are always the electrons in the outer shells. The protons and neutrons are safely tucked away in the middle of the atoms at their nuclei. So, the electrons in the outermost shell act as the "bumper" and determine how atoms interact with each other. We will discuss specifically how they interact shortly.

Remember these two points:

1. The number of protons in all atoms of a particular element is constant.

2. The number of protons in an atom = the number of electrons in that atom.

From this, we see that all atoms of a particular element have the same number of electrons. Because the electrons determine their chemical activity, all atoms of a particular element will react the same way.

Answers: 1. The period tells you how many shells of electrons there are. 2. The group tells you how many electrons are in the outermost shell.

In bumper cars, the outer rubber bumpers of the cars make contact first. The riders inside the cars should never contact each other.

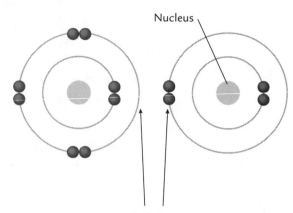

Nucleus

In atoms, the outer shell electrons make contact first. The protons and neutrons inside the nuclei never contact each other.

FIGURE 5.6 **Chemicals interact somewhat like bumper cars.** The electrons in the outer shells become the atoms' "bumpers" and determine the chemical's reactivity.

TIME TO TRY

Consider carbon in the Periodic Table.

What is its atomic number? _____

How many protons does it have? _____

How many electrons? _____

Once you see that carbon's atomic number is 6, you know that it has 6 protons, and so it also has 6 electrons. Now draw the electrons for carbon around its nucleus.

You should see that carbon has 4 electrons in its second (outer) shell, and these are the electrons that will interact with other electrons.

🔑 **The electrons in an atom's outermost shell determine its chemical reactivity.** ■

The Union: **Chemical Bonding**

As mentioned earlier, two or more atoms can join together through **chemical bonding** to form a molecule. Recall that the atom's electrons are arranged around the nucleus in one or more shells. These shells can have multiple subshells. The outermost of these subshells is called the **valence shell**. For simplicity, we will refer to this as the outermost shell. It can contain at most 2 electrons for helium, or 8 electrons for all other elements. If this outermost shell contains the maximum number of electrons, or is full, the atom is amazingly stable. It is said to be chemically **inert**—it will not easily react with other atoms. It is "happy," so to speak. All of the elements in the last column of the Periodic Table are inert.

On the other hand, atoms of elements in all of the other columns lack a full outermost shell. That means they are unstable and want to become stable. If it helps you remember this, think about life. When we are ful**filled**, we feel happy. If we are not happy, perhaps we feel something is missing from our lives, or maybe we feel we have lots of good to give and nobody to give it to. Now don't you feel sorry for those unfulfilled atoms?

An atom that does not have a full outermost shell of electrons is not stable, and it will react with other atoms to try to become stable. Unstable atoms can gain, lose, or share electrons with other unstable atoms until they become stable. That's how atoms interact. Let's explore this more deeply.

✔ **QUICK CHECK**

Under what circumstances is an atom stable? _____

_____

_____

Answer: An atom is stable when its outermost, or valence, shell is full, meaning it has 2 electrons for helium or 8 electrons for all other elements.

IONIC BONDING

When atoms become stable by losing and gaining electrons, electrons actually leave one atom's outermost shell and join the outermost shell of another atom. The atoms are now stable, meaning they each have a full outermost shell. However, gaining or losing electrons also changes the atoms in another way. Recall that atoms are normally electrically neutral—they have the same number of protons (+) and electrons (−). Once the electrons move, though, the atoms are no longer neutral because the protons and electrons are no longer balanced. An atom that gains an electron has one extra negative charge, and an atom that loses an electron is short one negative charge, making it positive.

All atoms that have gained or lost electrons carry an electrical charge and are called **ions**. These are designated with a $^+$ or $^-$ sign. For example, sodium tends to lose an electron and become a sodium ion, **Na$^+$**. Chlorine tends to gain an electron, becoming a chloride ion, **Cl$^-$**. Ions of opposite charges attract each other (*"Opposites attract"*). Whenever ions are formed, oppositely charged ions will join to form a compound that is electrically neutral. When they join, they form a strong **ionic bond**— ions form ionic bonds.

Atoms that gain or lose electrons form ions, and ions of opposite charges form ionic bonds. ■

TIME TO TRY

Let's see how ionic bonding works, using sodium and chlorine. Look at the Periodic Table and fill in the following information:

	Sodium (Na)	Chlorine (Cl)
Atomic number:	_____	_____
Number of protons:	_____	_____
Number of electrons:	_____	_____

You know from the Periodic Table that neither of these elements is stable—they are not in the last column of the Table, so their outermost shells are not full. How can they become stable? _____

_____

Draw the electrons around the nuclei of each of these atoms (the shells are drawn for you):

Sodium   Chlorine

Sodium, with atomic number 11, should have 2 electrons in its inner shell, 8 in the second, and a single electron in its outermost shell. Chlorine, atomic number 17, should have 2 electrons in its inner shell, 8 in its second shell, and 7 in its outermost shell. That is pretty convenient—sodium has one too many and chlorine is short one. Sodium will lose its electron to chlorine, producing two ions, Na^+ and Cl^-. Once the ions are formed, their opposite electrical charges will draw them together and they will form a strong ionic bond, creating a substance called *sodium chloride*. You know it better as table salt!

Now remember that magic from the Periodic Table: Sodium is in the first column, so it has 1 extra electron it wants to lose. As we saw earlier, all elements in that column have one electron in their outermost shell. Chlorine is in the next to last (stable) column. All of the elements in that column need only 1 electron to have a full outermost shell and be stable.

What would you predict about calcium? _____

What about oxygen? _____

Calcium is in the second column, so it has 2 electrons in an outermost shell that wants 8. It is not stable. Oxygen is two columns short of being stable, so it needs 2 more electrons to fill its outermost shell and be stable.

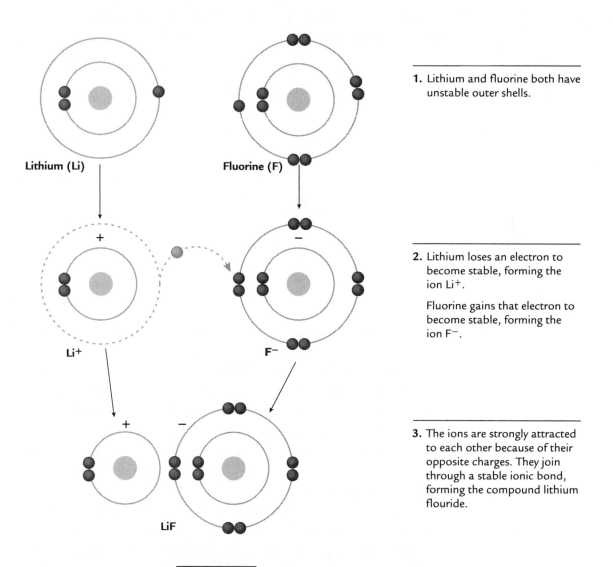

1. Lithium and fluorine both have unstable outer shells.

Lithium (Li)

Fluorine (F)

Li+

F−

LiF

2. Lithium loses an electron to become stable, forming the ion Li+.

   Fluorine gains that electron to become stable, forming the ion F−.

3. The ions are strongly attracted to each other because of their opposite charges. They join through a stable ionic bond, forming the compound lithium flouride.

FIGURE 5.7   **Ionic bonding.** Lithium will lose an electron to fluorine, forming two oppositely charged ions. These ions are then attracted to each other and form an ionic bond, producing the compound called *lithium fluoride*.

For review, **Figure 5.7** shows how an ionic bond forms between lithium and fluorine. These elements are in the same columns as sodium and chlorine, so the process is the same except these new elements have only two shells of electrons. But, as you now know, in chemical interactions, only the electrons in the outermost shells are important.

✔ **QUICK CHECK**

How does an ionic bond form? _____

_____

_____

COVALENT BONDING

I have an older brother. Growing up there were times, of course, when we, um, shall we say, disagreed? This often revolved around possession of some toy or other item. If you have siblings, you know how these squabbles usually ended—a parental voice from somewhere in the distance yelling for us to . . . *share.*

Apparently some atoms have learned that lesson as well. Let's consider two hydrogen atoms, each of which has a single electron. To form an ionic bond, one hydrogen atom would have to give up its electron and another atom would have to gain it. But which will gain and which will lose? (From my childhood I remember many long standoffs in which neither I nor my brother had any intention of giving up anything!) Neither; instead, both hydrogen atoms can share their electrons. By combining them, the electrons will orbit around both nuclei together, and both atoms will be stable as long as they stay together. This type of bond is called a **covalent bond**. Recall that the outermost shell is called the valence shell and it contains the electrons that are interacting. The atoms that are sharing electrons in order to have full valence shells are said to be covalent (*co-* as in together or cooperating; the electrons share a common valence shell). **Figure 5.8** illustrates the formation of a covalent bond. In general, elements that are closer to the right or left side of the Periodic Table are more likely to form ionic bonds, and those closer to the middle of the Table are more likely to form covalent bonds.

Answer: An ionic bond forms when atoms gain or lose electrons, forming oppositely charged ions that are then drawn together by their charges.

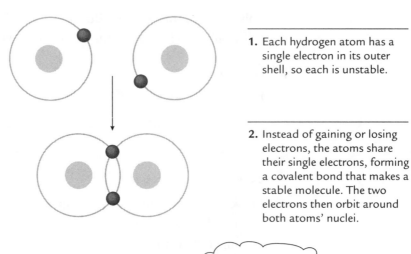

1. Each hydrogen atom has a single electron in its outer shell, so each is unstable.

2. Instead of gaining or losing electrons, the atoms share their single electrons, forming a covalent bond that makes a stable molecule. The two electrons then orbit around both atoms' nuclei.

Everyone is happy when they share.

FIGURE 5.8   **Covalent bonding.** In covalent bonding, instead of engaging in a tug of war, two atoms share the electrons in their outer shells to become stable. Here a covalent bond between two hydrogen atoms is shown.

✔ **QUICK CHECK**

How does a covalent bond form? _____

_____

HYDROGEN BONDING

Although there are many types of chemical bonds, we will look at just one more. A **hydrogen bond** is a weak bond that can form between the

Answer: A covalent bond forms when atoms share electrons to attain full outer-most shells and become stable.

hydrogen atoms in one molecule and some atoms in other molecules. Water is a classic example of this kind of bonding. Look at **Figure 5.9.** A water molecule has two hydrogen atoms and one oxygen atom. A hydrogen atom is tiny, with just 1 electron and 1 proton. An atom of oxygen is much bigger—it has 8 electrons, 8 protons, and 8 neutrons. Each hydrogen atom binds to oxygen by sharing its lone electron. Then the electron from hydrogen orbits around both nuclei—its own and that of oxygen. The electrons from both hydrogen atoms spend more time around the oxygen than around their own nuclei because the oxygen contains more positive-charged protons to attract the negative electrons.

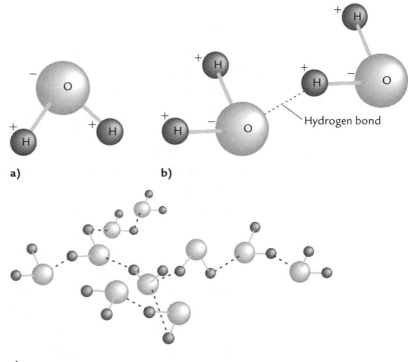

a)

b)

c)

FIGURE 5.9 **Hydrogen bonding. a)** A water molecule is polarized—the hydrogen atoms tend to be slightly more positive than the oxygen. **b)** A hydrogen bond forms between a slightly positive hydrogen atom and a slightly negative oxygen atom. **c)** Hydrogen bonds between water molecules give water many unique properties.

PICTURE THIS

Prove this to yourself with Figure 5.9a. Using your finger to represent the path of the shared electron, move your finger at a steady pace to trace the path around one hydrogen atom and move from there to pass around the oxygen atom. Your finger is near the oxygen longer. Although this simulation demonstrates that the electrons spend more time around the oxygen, it is the composition, rather than the size, of the oxygen and hydrogen atoms that determines this. Oxygen has 8 protons ($+$) to attract the hydrogen's electrons ($-$), but each hydrogen has only a single proton to attract them. Electrons are more strongly attracted to the location with the most protons.

In addition, oxygen only shares 2 of its electrons, so it always has 6 around it, even without the shared electrons. So at any given time, there are more electrons around the oxygen than around the hydrogen. As a result, there is a slight imbalance in the electrical charge of the water molecule—the oxygen tends to be a bit more negative than the hydrogens because the electrons hang out there more. This slight charge imbalance is called **polarity**.

Polar molecules can form weak hydrogen bonds, as shown in **Figure 5.9b**. The slightly more positive hydrogen is attracted to the slightly more negative oxygen in an adjacent water molecule, so they form a weak bond. This is the hydrogen bond. It is kind of like striking up a conversation with a stranger in the checkout line at the grocery store. You happen to be standing close so perhaps you chat briefly, but you are not going to become lifelong buddies from that short, superficial exchange. Although they are weak, hydrogen bonds are important. They help shape many important molecules, such as proteins, your individual hairs (straight or curly), and your DNA.

Water has lots of hydrogen bonds, as shown in **Figure 5.9c**, giving it many unique characteristics. For example, ice floats because the hydrogen bonds cause the water molecules to spread out more in the solid form than they do in the liquid form. Water also has a high boiling point because of its hydrogen bonds.

TIME TO TRY

Let's examine the compatibility of polar and nonpolar substances.

1. Take any clear container, preferably one that can be closed. A plastic baggie will work. Fill it about a third full with water.

2. Add to that about half as much cooking oil. Try to get them to mix and observe what happens. _____

_____

3. Add some food coloring to the container and shake it well to mix.

4. Is the food coloring polar or nonpolar (*Hint: Which layer is it in?*)

_____

5. Now find three harmless liquids and mix each one with water. Avoid household cleaners that may be caustic or react with each other if mixed.

Liquid Tested   **Polar or Nonpolar**

_____   _____

_____   _____

_____   _____

You likely found that your liquids, unless they were other oils, were polar, because they mixed with the polar water layer. Water is often used to make **solutions** because so many substances will dissolve in it. In a solution, one substance—the **solute**, dissolves in another substance—the **solvent**. Water is considered a universal solvent, and most chemical reactions in the human body occur in water. Water readily dissolves polar molecules, but it causes nonpolar molecules to bunch together.

Water also has high *adhesion*, meaning it sticks to surfaces very well, and high *cohesion*, meaning its molecules stick to each other. Let's demonstrate that.

TIME TO TRY

You will need two pennies, alcohol, water, a dropper (or you can carefully use your finger), and a paper towel.

1. Place the pennies on a paper towel and examine them. Estimate how many drops of liquid you can put on a penny before it will spill over. _____ drops

2. Start with the alcohol. Using the dropper or your finger, carefully place drops of the alcohol on the surface of the penny, counting each drop, until it overflows. How many drops of alcohol fit on the penny? _____

3. Now use water and repeat this on the other penny. How many drops of water fit on the penny? _____

You should have been able to pile more water than alcohol on the penny because the water molecules stick to the penny and to each other much more than the alcohol molecules do. Because most beverages have a high water content, this property also allows you to fill a glass slightly higher than the rim (just don't try to put a lid on it or pick it up!)

✔ **QUICK CHECK**

Explain what is meant by *polar* molecule. _____

_____

MOLECULES AND COMPOUNDS

As mentioned earlier, when two or more atoms bind together, they form a **molecule**. If the atoms are from the same element, they form a molecule of that element; for example, O_2 is a molecule of oxygen. If the atoms are from different elements, the substance formed is called a **compound**. Water is a compound because it contains two elements.

A molecule is described by a **molecular formula** that tells you what the molecule is made of. This formula includes the letter symbols for the

Answer: A polar molecule is one in which there is an uneven charge distribution across the molecule, resulting in slightly positive and slightly negative charges for different parts of the molecule.

elements and the number of atoms of each element that are present in the molecule. The numbers are always in the subscript position. For example,

H_2O = water $\qquad\qquad\qquad$ CO_2 = carbon dioxide

O_2 = oxygen $\qquad\qquad\qquad\;\;$ CO = carbon monoxide

Note that the only difference between carbon dioxide and carbon monoxide is one atom of oxygen. We make carbon dioxide in our body and exhale it with every breath, whereas carbon monoxide is a deadly poison.

The molecular formula gives us limited information. It tells us how many pieces are in a molecule, but not how they are hooked together. For that, we can consult the **structural formula**, which is a simplified drawing of how the molecule is built. Lines in a structural formula represent chemical bonds (see Figure 5.9). Now look at **Figure 5.10**. This shows the structural formulae for three sugars: glucose, galactose, and fructose. Glucose and galactose are quite similar, so the differences are highlighted. Fructose, also called fruit sugar, has an obviously different appearance.

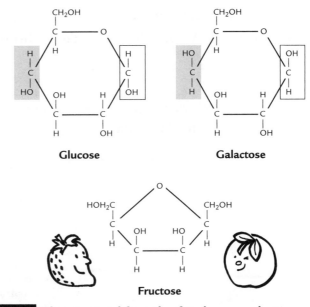

Glucose

Galactose

Fructose

FIGURE 5.10 **The structural formulae for glucose, galactose, and fructose.** Although they all have the same molecular formula, the highlighted areas show the differences between glucose and galactose, and fructose is more obviously different.

TIME TO TRY

Look carefully at Figure 5.10 and fill in the following information.

	Glucose	Galactose	Fructose
Number of carbon atoms:	_____	_____	_____
Number of hydrogen atoms:	_____	_____	_____
Number of oxygen atoms:	_____	_____	_____
Molecular formula:	C _ H _ O _	C _ H _ O _	C _ H _ O _

(*Hint: Use the numbers you wrote for each element above.*)

You can see that all three of these sugars have the same molecular formula: $C_6H_{12}O_6$. The structural formula provides more detailed information and is often more useful than the molecular formula. However, the molecular formula is the most common method of describing molecules and compounds.

✔ **QUICK CHECK**

1. What is the basic difference between a molecule and a compound?

_____

_____

2. What is the difference between a molecular formula and a structural formula? _____

_____

Answers: 1. A molecule is formed whenever two or more atoms join; if they are from different elements, the substance formed is a compound. 2. A molecular formula tells you how many atoms of each element are in a molecule; the structural formula shows how the atoms are connected.

Double Bubble, Toil and Trouble: **Electrolytes and pH**

Molecules are categorized in many ways. Let's explore a group called **electrolytes.** Recall that ions are atoms with electrical charges. Any compound that releases ions in water is an electrolyte, and there are three types:

- **Acids** release H^+ in water.

- **Bases** form ions that can bind with H^+.

- **Salts** are produced when acids and bases combine with each other.

ACIDS

Let's start by looking at acids. How does hydrogen become the ion H^+?

Recall that positive ions form when electrons are lost. H^+ is an atom of hydrogen minus its electron—it is primarily a proton. Thus, acids are called *proton donors.* Acids in the home might include vinegar (acetic acid), carbonated soft drinks (phosphoric and citric acids), and citrus fruit (citric acid). Acids in your body include hydrochloric acid (HCl) in your stomach and lactic acid that can accumulate in your muscles.

BASES

Bases form ions that bind H^+, so they are *proton acceptors.* Bases in the home include ammonia and bleach. Bases in your body include compounds that release bicarbonate ions, such as sodium bicarbonate (also in baking soda). Bases minimize the damaging effects of acids in your body. For example, the hydrochloric acid from your stomach is tamed by bicarbonate in your intestine.

SALTS

When acids and bases react together, they produce salts and water. As electrolytes, salts release ions in water, providing many important ions, such as calcium, potassium, sodium, magnesium, and chloride. Salts also help maintain our water balance, which in turn affects our blood volume, blood pressure, and normal cell functions.

WHY SHOULD I CARE?

Ions from electrolytes play such critical roles in keeping us alive that in the medical field the term "electrolyte" is often used to refer to the ions themselves, rather than the compounds that produce them. Maintaining proper amounts of these ions is essential for normal body functions. Calcium, for example, is required for nerve communication, muscle contraction, and strong bones and teeth. Maintaining the proper *fluid-electrolyte balance* is essential for homeostasis.

ACID-BASE BALANCE AND pH

Many physiological processes depend on certain amounts of acids and bases—the *acid-base balance.* Like the fluid-electrolyte balance, acid-base balance is essential for our survival. Acids release H^+, and the more H^+ that is present, the more acidic the solution is. Bases bind H^+, so the more base that is present, the less H^+ there is. The amount of acid and base in a solution is measured in pH units. The **pH** refers to how much H^+ is present. The **pH scale** ranges from 0 to 14 (**Figure 5.11**). Pure water has equal amounts of H^+ and OH^- (hydroxide ion, which acts as a base). Because of this, pure water has a neutral pH of 7.0. A pH below 7.0 is acidic; a value above 7.0 is basic or *alkaline* (another term for basic). Each one-unit change in pH is actually a ten-fold change because

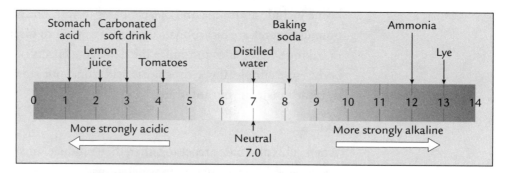

FIGURE 5.11 **The pH scale measures the H^+ concentration in solutions.** The scale has values from 0 to 14; 7.0 is neutral, with equal amounts of acid and base. The pH values of some familiar solutions are shown.

each pH unit represents a power of 10. The difference between a pH of 1 and a pH of 2 is 10^1 (10) and the difference between a pH of 1 and a pH of 3 is 10^2 (100). H^+ concentration and pH are inversely related—as H^+ concentration goes up, pH drops, and vice versa. In other words, the stronger the acid, the lower the pH, and the stronger the base, the higher the pH.

✔ **QUICK CHECK**

Gastric juice—the liquid in your stomach—has a pH around 1.5. Coffee's pH is about 5.0, and most colas are around 3.0. Which of these is the most acidic, and which is the least? _____

I'm In with the In Crowd: **Inorganic Compounds**

Chemical compounds are classified as organic or inorganic. **Organic compounds** contain both carbon and hydrogen. All others are classified as **inorganic compounds,** which may contain C or H, neither of them, but not both. *Water* (H_2O) is our most abundant inorganic compound. It's the major component of all body fluids. It both bathes and fills our cells, and transports substances to and from them. Most chemical reactions occur in water, and it allows us to lose excess heat through sweating. Water is, arguably, our most important nutrient.

We take in *oxygen* (O_2) by breathing, and it enters our blood through our lungs. Most oxygen travels in our blood and is released where it is needed. Oxygen allows us to get the most energy from foods we eat by freeing the energy in food molecules. This energy fuels our cells' activities. While using oxygen to extract energy, cells produce *carbon dioxide* (CO_2) as a waste product. This CO_2 moves through your blood to your lungs where it is exhaled.

Answer: Gastric juice is the most acidic, colas are next, and coffee is the least acidic.

TIME TO TRY

The pigments in red cabbage can act as a pH indicator. Let's try it. You need red cabbage, boiling water, a clear glass container, a drinking straw, and baking soda.

1. Chop 1 cup of red cabbage into enough boiling water to cover it. Let this stand for 15 minutes.

2. Put about 1/4 cup of the water—the indicator—in your clear container. Its current color indicates a near-neutral pH of 7.0. *What color is the indicator for a neutral pH?* _____
   With an acidic pH, the indicator turns more red; with a basic pH, it turns more bluish-green.

3. Use the straw to blow bubbles in the indicator. Blow until the color changes. (Get comfortable—this could take awhile.) *What color is it turning?* _____ *What does this say about the pH?* _____

4. Now add baking soda to the indicator and stir it. Repeat this until the indicator returns to the neutral color.

5. Add double the amount of baking soda you first put in and stir. *What color is the solution now?* _____ *Is baking soda an acid or a base?* _____

Blowing bubbles (CO_2) into the water creates carbonic acid, and the indicator should turn redder. Baking soda is a base, binding the H^+ from the carbonic acid, and the indicator should return to the neutral color, then move into the basic range as you add more. The main reason we exhale is to clear CO_2 from the body. If we don't clear enough CO_2, too much carbonic acid forms and our pH drops. Fortunately, our bodies also produce bicarbonate ion to help neutralize the H^+ from the carbonic acid.

Is it **Organic?**

The term *organic* is used today to describe artwork, home décor, and how food is grown, among other things. In chemistry, though, it has a very precise meaning. By definition, **organic compounds** contain both carbon and hydrogen. They must have these two elements, and they usually have others as well. Most of your body is made of organic molecules. That, Earthling, is what is meant by saying we are a carbon-based life form. The main categories of organic compounds are

- carbohydrates (sugars and starches),

- proteins,

- lipids (which include fats and steroids), and

- nucleic acids (DNA, which is your genetic material, and RNA, which assists DNA).

✔ **QUICK CHECK**

What is the difference between organic and inorganic molecules?

_____

CARBOHYDRATES

Carbohydrates include sugars and starches. All carbohydrates contain C, H, and O, with hydrogen and oxygen present in a 2:1 ratio. That means carbohydrates contain twice as much hydrogen as oxygen. For example, the molecular formula for glucose is $C_6H_{12}O_6$. Sucrose (table sugar) is $C_{12}H_{22}O_{11}$. Notice that the 2:1 rule does not involve carbon. *What common small molecule can you think of that also has this ratio?*

_____

Hopefully you got that: it's good ol' H_2O—water.

PICTURE THIS

It's a hot summer day. You've been sweating heavily and feel over-heated and weak. You go inside to cool off and quench your thirst. Heavy sweating caused you to lose more water than you should.

What term do we use for that situation? *We would say that you are*

de__ __ __ __ __ __ __.

You drink some water and feel refreshed because you are

re __ __ __ __ __ __ __.

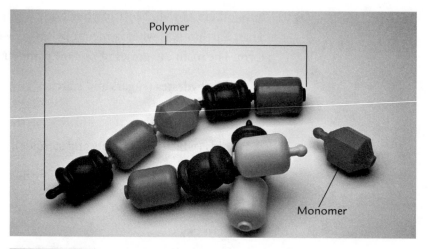

FIGURE 5.12 **Monomers are like children's pop beads.** Each bead is one unit, or a monomer. Monomers can be linked together to form larger structures called polymers.

The term "hydrate" refers to water. Now look again at the name of this category of compounds—*carbohydrate*. It literally means carbon that is hydrated—carbon with water. That is why we expect carbohydrates to have the 2:1 ratio of hydrogen to oxygen.

Many organic compounds are organized from small units called **monomers** (*mono* means *one*). Monomers are the simplest form of each category of compounds, and they can link together into larger, more complex compounds called **polymers** (*poly* means *many*). To understand this, consider children's pop beads (**Figure 5.12**). Each "bead" is like a monomer, and when monomers join they form a polymer. Our foods contain polymers, but digestion breaks them down to monomers so they can enter our blood for delivery to our cells. Once in the cells, monomers are used directly or formed into various polymers for your body's needs.

The major carbohydrate monomer is $C_6H_{12}O_6$ (**Figure 5.13**), which is a sugar. Another term for sugar is *saccharide.* Because this is a sugar monomer, $C_6H_{12}O_6$ is also called a **monosaccharide**—it is a single sugar unit. Our most common monosaccharide is *glucose.* Two monosaccharides linked together make a **disaccharide,** such as table sugar, or *sucrose.* If more monosaccharides join, the product is a **polysaccharide,** also known as a complex carbohydrate. *Starch* is a polysaccharide, as is

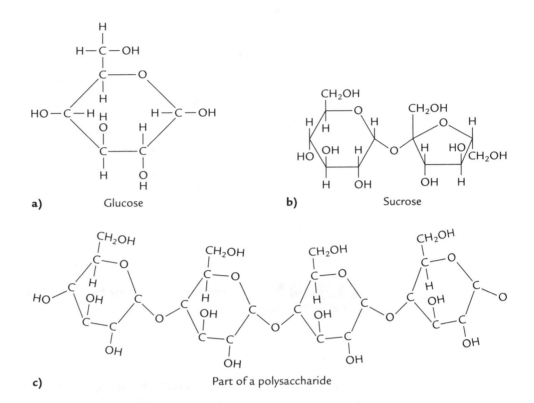

a) Glucose

b) Sucrose

c) Part of a polysaccharide

FIGURE 5.13 **Carbohydrates. a)** A monosaccharide, such as glucose, is a carbohydrate monomer. **b)** Two monosaccharides linked together form a disaccharide, such as sucrose. **c)** More than two monosaccharides unite to form a polysaccharide. Part of a polysaccharide is shown here—note the repeating monomer units.

glycogen, the form in which our cells store glucose. Carbohydrates provide energy and building materials. Dietary sources of carbohydrates include sweets and starches and food groups like fruits, vegetables, and grains.

✔ **QUICK CHECK**

What are the monomers of carbohydrates? _____

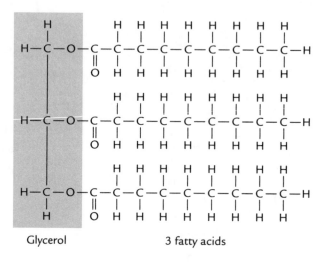

Glycerol   3 fatty acids

FIGURE 5.14 **A triglyceride (fat) has three fatty acids linked together by glycerol.**

LIPIDS

Lipids, which include fats and steroids, among others, don't dissolve in water. Our most familiar lipid is **fat**. Like carbohydrates, fats contain C, H, and O, but the ratio of hydrogen to oxygen is much higher in fat. Consider the formula for one fat: $C_{57}H_{110}O_6$. It has 110 hydrogens and only 6 oxygens! A single molecule of fat is also called a **triglyceride,** and it contains three carbon chains called *fatty acids* linked together by a small compound called glycerol (**Figure 5.14**). These compounds are the monomers. Dietary sources of fat include many meats, cheese, whole milk, oils, and butter. Much of our pleasure from food comes from the smooth texture, or "mouth feel," that fats contribute.

In our bodies, fat is a major energy source; in fact, we store extra energy (Calories) as fat. It insulates us (ask any walrus how important blubber is in frigid water!), cushions our organs, and some of our vitamins need fat to enter our bodies. Fat is not the only type of lipid, though. *Phospholipids* are the main component of cell membranes. *Steroids* include cholesterol, which is an essential part of our cell membranes, and some hormones, like testosterone and estrogen.

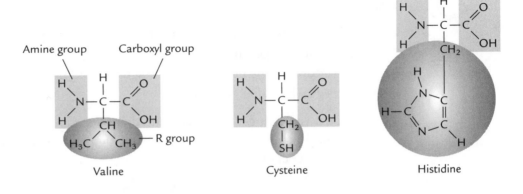

Amine group   Carboxyl group   R group

Valine   Cysteine   Histidine

FIGURE 5.15 **Amino acid.** An amino acid contains an amine group, a carboxyl group, and a variable area called "R." Compare these amino acids and you will see that the R area is the only place that they differ.

PROTEINS

Like carbohydrates and fats, **protein** contains C, H, and O, but it also has nitrogen (N). Protein's monomers are **amino acids** (**Figure 5.15**). Each amino acid has an *amine* group ($^-NH_2$) and another part called a *carboxyl* group (^-COOH), both attached to a central carbon. The carboxyl group allows the compound to act as an acid—hence the name amino acid. Another part, the *R group*, makes each amino acid distinct.

Each protein contains amino acids linked together in a specific sequence, called the *primary structure*. Think of it as a string of assorted beads making a necklace—if you change any bead or their order, you make a different necklace. Similarly, the specific order of amino acids determines what the protein is and how it functions. A protein's primary structure is determined by genes—your DNA. The chain of amino acids is usually twisted and folded into a unique three-dimensional shape by chemical bonds, and some proteins even contain more than one chain twisted and folded together. This sounds complicated, but it is very important—the protein's specific 3-D shape determines how it works, and any change in that shape can make it nonfunctional.

TIME TO TRY

Time to head to the kitchen for a snack. Crack an egg into a pan and describe the egg white. _____

Now turn on the heat, cook the egg until it is "done," and describe the egg white after cooking. _____

Oops! You changed your mind and don't want an egg after all. Put it back in the refrigerator to cool. Does it go back to its original state after cooling? _____

Egg white contains a protein called *albumin*. It comes out of the shell as a clear, thick liquid, but when heated it becomes a white, rubbery solid. A protein's shape is maintained by hydrogen bonds cross-linking parts of the amino acid chain to each other. These bonds can be broken by heat. If that happens, the protein is *denatured*—it becomes disorganized and loses its unique shape and thus its special properties. Denaturation can't be reversed—you can't uncook an egg white. In cells, if proteins are denatured they may be nonfunctional.

Dietary sources of protein include meat, fish, poultry, dairy products, nuts, and legumes. Proteins provide structural materials for growth and repair. Many hormones are built from proteins, as are cell receptors that allow materials to enter, antibodies that keep us healthy, and enzymes. **Enzymes** are especially important for body processes—they allow chemical reactions to occur. All enzymes are proteins, and most chemical reactions in the body require specific enzymes. They assist other chemicals, acting a bit like tour guides ensuring that the chemicals get where they need to be and do what they are supposed to. They make it easier for reactions to occur. Without enzymes, very little work could occur in our cells—almost no energy production, synthesis, or repair. Our metabolic processes, and our existence, would grind to a halt.

✔ **QUICK CHECK**

How is the term amino acid related to proteins? _____

_____

Answer: Amino acids are the monomers that form proteins; their name reflects that they contain an amine group and an acidic carboxyl group.

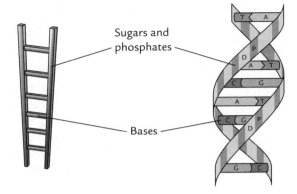

FIGURE 5.16   **Structure of DNA.** The chemical structure of DNA is like a ladder, with sugars (D) and phosphates (P) forming the uprights and organic bases (A, T, C, and G) forming the steps.

NUCLEIC ACIDS

The last organic compounds to explore are **nucleic acids—DNA** and **RNA.** Each nucleic acid contains C, H, O, N, and phosphorus (P). Nucleic acids are large, complex compounds, and **nucleotides** are their monomers. Each nucleotide contains a sugar, a phosphate group, and an organic base. Nucleic acids are named for the sugar they contain.

PICTURE THIS

Imagine a ladder. The uprights (hand rails) represent alternating sugars and phosphates, with the "steps" attached to the sugars. These steps are paired organic bases—adenine (A), thymine (T), guanine (G), and cytosine (C). They match up specifically—adenine and thymine are always together, and guanine is always on a step with cytosine. Now, imagine twisting your ladder into a spiral staircase. You just made a double helix, the classical shape of DNA.

Deoxyribonucleic acid (DNA) contains the sugar deoxyribose; *ribonucleic acid (RNA)* instead contains ribose. Nucleic acids consist of many nucleotides linked together in a long chain and folded into special shapes.

DNA is your genetic material, and it is organized into chromosomes (**Figure 5.16**). DNA contains stacked pairs of nucleotides. The organic bases store information in the form of a molecular code that provides instructions for building your proteins. A span of DNA with instructions for a single polypeptide is a *gene.* These instructions are what you inherit from your parents—half from mom and half from dad.

RNA's three major types are *messenger RNA (mRNA), ribosomal RNA (rRNA),* and *tRNA (transfer RNA).* RNA's job is to carry out the DNA's instructions. Think of DNA as the blueprints for building a protein. RNA, then, is like the workers that do the building. We'll discuss the specific roles of nucleic acids more in Chapter 6.

Love is Just a **Chemical Reaction**

All activities that occur within our bodies, including the functioning of our brains and affairs of our hearts, start with chemical reactions. When chemicals react with each other, bonds are formed or broken to produce new chemical combinations or to release ions. Energy is stored, energy is released, and energy is used in chemical reactions. Many aspects of physiology involve amazingly complex reactions that are meticulously controlled, whereas others are quite simple.

Chemical reactions are written in the form of **chemical equations**, but instead of using an equals sign, we use an arrow. The substances to the left of the arrow are the **reactants**—the things that react together. The items to the right of the arrow are the **products**—the end result of the reaction. The arrow means "produces" or "yields." For example, $Na^+ + Cl^- \rightarrow NaCl$ would be read "sodium ion plus chloride ion produces sodium chloride." Many reactions are **reversible**, meaning they can go in either direction. You can combine the ions we just mentioned to make salt, but salt can also break down to produce the ions. Reversible reactions are often indicated with a special double arrow symbol: $\rightleftharpoons$. Let's examine three basic types of reactions: synthesis, decomposition, and exchange.

Synthesis reactions are reactions that build. Another term for these is **anabolic reactions**. With that in mind, think about why athletes sometimes engage in illicit use of *anabolic* steroids—to *build* their muscles and strength. In synthesis reactions, two or more atoms or molecules combine to make a larger molecule. For example, proteins combine to build muscles. Small sugars combine to build large molecules of

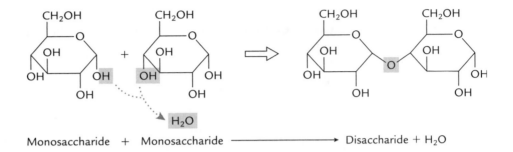

Monosaccharide + Monosaccharide $\longrightarrow$ Disaccharide + H_2O

FIGURE 5.17 **Two monosaccharides can join by dehydration synthesis.** In this process, a hydrogen is removed from one monosaccharide, and an oxygen and a hydrogen are removed from the other one. Once removed, they form water and the monosaccharides attach at the removal site, forming a disaccharide.

starch. This is done by forming new chemical bonds to hold the pieces together. Synthesis reactions are especially important in humans for growth and repair processes.

The most common type of synthesis reaction is **dehydration synthesis.** Think about that term. What does "dehydration" mean? _____

_____

In dehydration synthesis, adjacent monomers are connected by removing water and linking them together at the removal sites. Specifically, a hydrogen atom is removed from one monomer, and both a hydrogen and an oxygen are removed from an adjacent momomer (**Figure 5.17**). A quick glance shows us that the removed atoms make water. The monomers then join where these atoms were removed. Dehydration synthesis is how most organic monomers form polymers.

Decomposition reactions are the opposite of synthesis reactions. During decomposition reactions, larger structures are broken down into smaller parts. These are also called **catabolic reactions.** For example, starch is broken down into smaller sugars. Salt is broken down to ions. These reactions are done by breaking chemical bonds. Decomposition reactions are especially important in humans for digesting food and getting energy.

For organic molecules, the most common type of decomposition is **hydrolysis**. Think about that term and refer to the word root and suffix tables in Chapter 3.

What does hydro *mean?* _____

What does lysis *mean?* _____

Hydro means water; *lysis* means destroy or rupture. Decomposition is the opposite of synthesis, thus hydrolysisis is the opposite of dehydration synthesis. Hydrolysis uses water to *break* bonds. Bonds are split and water is put back in at the same places where it was removed during dehydration synthesis—one hydrogen from the water joins the already present oxygen, and the other hydrogen and oxygen from the water join the other broken bond. This is the basic process that breaks down food into usable units.

Exchange reactions involve swapping pieces. Two or more molecules split apart and then recombine in a new way; for example, AB and CD separate, then recombine to form AC and BD. Exchange reactions allow the body to receive and store chemicals in one form and then reuse them for multiple purposes. Here is an example of an exchange reaction:

$$HCl + NaOH \rightarrow NaCl + H_2O$$

Notice that both of the original compounds separate and swap parts. The final products have the same atoms, but they are rearranged.

WHY SHOULD I CARE?

Physiology is the study of how the body and all of its parts work. Although it may be hard to realize it, all of the work done by the body—collectively referred to as *metabolism*—involves chemical reactions. We eat and breathe to bring in the necessary chemicals, then the body uses those molecules in an amazing array of chemical reactions that allow us to do virtually everything we do. All body processes rely on synthesis, decomposition, and exchange reactions. When you begin studying specific chemical processes in the body, it may help to understand them if you think of what the outcome should be—are the reactions building, breaking down, or swapping? If you know the goal, you can more easily understand the reactions.

✔ **QUICK CHECK**

What are the three basic types of chemical reactions? _____

_____

The Big Bang: **Chemistry and Energy**

In Chapter 4, we briefly discussed the importance of energy, which is the ability to do work. Because our energy comes from what we eat (chemicals) it's appropriate to end with this topic. Energy comes in many forms, such as mechanical, electric, solar, and nuclear. All types can be categorized as either kinetic or potential. **Kinetic energy** is energy that's in use—it's the energy of action, or the energy used to do work. **Potential energy** is energy waiting to be used—it's not doing work at the moment but has the *potential* to do work. It's like stored energy.

We previously defined *metabolism* as all the work going on in your body. Even when you're asleep, your body is very busy—your heart beats, you breathe, your brain controls the processes that keep you alive, for example. Such activities require energy, and you need a supply available at all times. Energy can't be created or destroyed—it can only change from one form to another. That means we have to get our energy from somewhere, but we can use only chemical energy. Recall that we rely on plants to change solar energy (sunlight) to chemical energy through photosynthesis, creating carbohydrates that store this energy. We eat the plants, or animals that ate them, to get chemical energy. All foods store energy in the chemical bonds that hold them together.

To use energy from our foods, we extract it from our foods then store it in a molecule called **adenosine triphosphate,** or **ATP.** All living organisms rely on ATP for their energy needs—it's universal. ATP is often

Answer: The three basic chemical reactions are synthesis, decomposition, and exchange.

referred to as our energy currency—to do work, we have to "spend" ATP. We do this by breaking a bond in the ATP and releasing the energy we stored there. We need a lot of ATP at all times, and we need to constantly make more to keep going. We do this primarily through a complex series of chemical reactions collectively called *cellular respiration.*

It's appropriate that chemistry occupies the first three levels of the biological hierarchy of organization. Chemistry is the foundation of all living organisms.

Our existence requires both organic and inorganic substances. These chemicals are constantly reacting, forming new bonds and breaking old ones, and they account for virtually everything you do and you are. Atoms and molecules are the very stuff of life, and the reactions between them are the processes of life. I hope now you can see that to achieve success in anatomy and physiology, chemistry truly does matter.

Final Stretch!

Now that you have finished reading this chapter, it is time to stretch your brain a bit and check how much you learned. For online tests, tutorials, animations, activities, web links, and an ebook, visit the *Get Ready for A&P* companion website.

RUNNING WORDS

At the end of each chapter, be sure you have learned the language. Here are the terms introduced in this chapter with which you should be familiar. Write them in a notebook or enter them in your computer. Define them in your own words, then go back through the chapter to check your meaning, correcting as needed. Also try to list examples when appropriate.

Matter	Proton
Mass	Neutron
Weight	Electron
Volume	Nucleus
Element	Orbital
Atom	Shell
Molecule	Atomic number
Macromolecule	Atomic mass/weight
Subatomic particle	Atomic mass unit (u)

Isotope

Periodic Table of
 Elements

Period

Group

Chemical bonding

Valence shell

Inert

Ion

Ionic bond

Covalent bond

Hydrogen bond

Polarity

Solution

Solute

Solvent

Compound

Molecular formula

Structural formula

Electrolyte

Acid

Base

Salt

pH

pH scale

Organic compound

Inorganic compound

Carbohydrate

Monomer

Polymer

Monosaccharide

Disaccharide

Polysaccharide

Lipid

Fat

Triglyceride

Protein

Amino acid

Enzyme

Nucleic acid

DNA

RNA

Nucleotide

Chemical equation

Reactant

Product

Reversible reaction

Synthesis reaction

Anabolic reaction

Dehydration
 synthesis

Decomposition
 reaction

Catabolic reaction

Hydrolysis

Exchange reaction

Kinetic energy

Potential energy

Adenosine
 triphosphate (ATP)

WHAT DID YOU LEARN?

**PART A: PROVIDE THE MISSING INFORMATION.
CONSULT THE PERIODIC TABLE, FIGURE 5.3.**

	Potassium	Iodine	Oxygen	Neon
Chemical symbol	_____	_____	_____	_____
Atomic number	_____	_____	_____	_____
Atomic weight	_____	_____	_____	_____
Number of protons	_____	_____	_____	_____
Number of electrons	_____	_____	_____	_____
Number of neutrons	_____	_____	_____	_____
Number of electrons in outermost shell	_____	_____	_____	_____

PART B: ANSWER THE FOLLOWING QUESTIONS.

1. Which subatomic particles are always in the nucleus? _____

2. Which subatomic particles are not included in the atomic weight? _____

3. What are the four main elements that make up most of the human body?

_____

_____

4. In the Periodic Table, what does the row (period) in which an element is positioned tell you about that element? _____

5. What does the element's column in the Periodic Table tell you? _____

_____

6. A calcium ion has a +2 electrical charge. How did this ion form to give it that charge? _____

7. What is the basic difference between an ionic bond and a covalent bond? _____

_____

8. Which gives you the most information: a molecular formula or a structural formula?

_____

9. When you eat, the food is converted into small, simple molecules that can be absorbed into your blood. What type of chemical reactions does that involve?

_____

10. If you get a paper cut, what type of chemical reactions will allow your skin to repair itself? _____

11. What elements are always found in organic molecules? _____

_____

12. For each of the following, is the compound organic (O) or inorganic (I)?

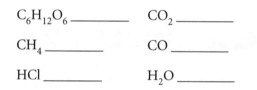

$C_6H_{12}O_6$ _____ CO_2 _____

CH_4 _____ CO _____

HCl _____ H_2O _____

13. What are the monomers of carbohydrates and of proteins? _____

14. Enzymes are examples of which type of organic compound? _____

_____

15. What molecule do we use directly to supply energy for our cells' work? _____

_____

6 Cell Biology

Life's Little Factories

When you complete this chapter, you should be able to:

■ Explain the Cell Theory.

■ Distinguish between prokaryotic and eukaryotic cells.

■ Describe the structures and functions of the cell membrane and cell organelles.

■ Explain various movement processes that occur in cells.

■ Describe the complete cell cycle and the basics of cell reproduction.

Your Starting Point

Answer the following questions to assess your knowledge about cells.

1. Give an example of a prokaryotic cell. _____

2. The liquid located between a cell's membrane and its nucleus is called

_____.

3. Another word for *single-celled* is _____.

4. What types of molecules make up the cell membrane? _____

_____

5. What is the function of a ribosome? _____

_____

6. Where in a cell would you find the genes that determine your eye color?

_____

7. What kind of molecules can easily pass through a cell membrane?

_____

8. What is *diffusion*? _____

_____

9. What moves during *osmosis*? _____

_____

10. What happens during *mitosis*? _____

_____

Answers: 1. Bacteria. 2. Cytoplasm or cytosol. 3. Unicellular. 4. Phospholipids, proteins, cholesterol, and some carbohydrates. 5. Build proteins. 6. In the DNA in the nucleus. 7. Lipid-soluble molecules. 8. Movement of molecules from an area of higher concentration to an area of lower concentration. 9. Water. 10. The nuclear contents divide.

Good Things Come in Small Packages: **Cell Theory**

In the last chapter, we explored chemistry and the first few levels of the Biological Hierarchy of Organization: atoms, molecules, and macro-molecules. In this chapter, we continue our climb up the ladder (**Figure 6.1**) as we explore the next two levels: organelles and cells, which are where most of the chemistry occurs in our bodies. This chapter's focus is **cytology**, which is the study of cells (*cyto-* = cell, *-ology* = study of).

In the 1600s, while examining slices of cork with a microscope, English naturalist Robert Hooke noticed that the cork was made of tiny chambers that reminded him of the cells in which monks lived in a monastery. From this observation, he coined the term **cell,** from the Latin word *cella,* meaning "storeroom" or "small container." Since then, vast amounts of cellular research have led to and continue to support a set of conclusions that are collectively referred to as the **Cell Theory**. Most sources acknowledge at least three major parts to this theory and some list as many as seven, so don't be surprised to see variation—all of these ideas are closely related. We shall consider the following five main principles:

1. All organisms are composed of one or more cells.

2. Cells are the basic structural and functional units of life.

3. All vital functions of an organism occur within cells.

4. All cells come from preexisting cells.

5. Cells contain hereditary information that regulates cell functions and is passed from generation to generation.

All organisms are composed of one or more cells. Some organisms are merely a single cell and are called **unicellular** organisms (*uni* = one). Organisms made of two or more cells are called **multicellular**. Whether the organism is a bacterium, fungus, plant, or animal (like you), it is made of cells, and the cells of all these organisms show tremendous diversity. Cells come in many different shapes, sizes, and types. In fact, we humans have over 200 different types of cells in our bodies (**Figure 6.2**). Red blood cells, or erythrocytes, are among our tiniest cells, measuring only about $2\,\mu$m thick and $7\,\mu$m in diameter. (Recall that a

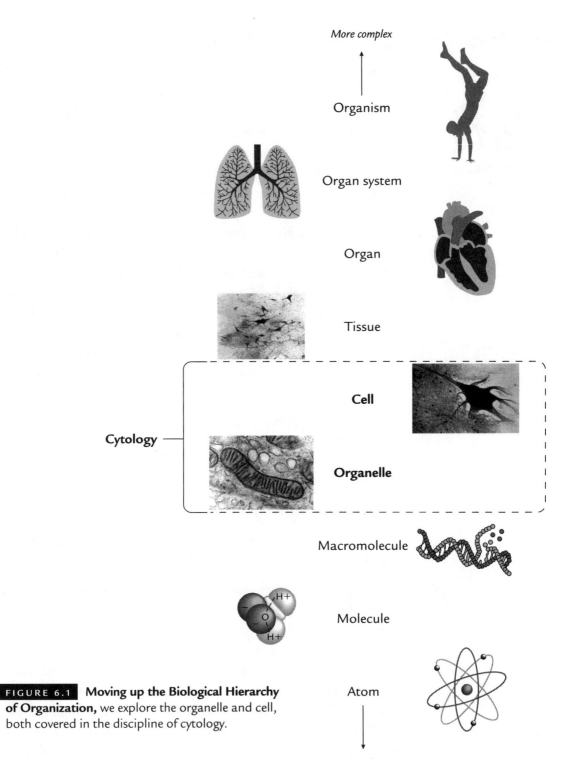

More complex

Organism

Organ system

Organ

Tissue

Cell

Cytology

Organelle

Macromolecule

Molecule

Atom

Simpler

FIGURE 6.1 **Moving up the Biological Hierarchy of Organization,** we explore the organelle and cell, both covered in the discipline of cytology.

micrometer is only 1/1000 of a millimeter.) At the other end of the spectrum, a single neuron (nerve cell) may be up to a meter long! Sperm are minute compared to the ova that they fertilize. Yet, all of the diverse cells found in all organisms share striking similarities.

Cells are the basic structural and functional units of life. Cells are certainly the basic structural units—all organisms are built from them. But cells are also the functional units of life. Housed within its single cell, a unicellular organism has all of the structures and the processes necessary to keep itself alive. Similarly, each cell in a multicellular organism, such as you or me, is an independent living unit capable of maintaining itself. It takes in nutrients, uses them to make the molecules it needs to function, and harnesses energy for doing its work. While going about its daily business, the cell generates and also rids itself of waste. And, in most cases, it reproduces.

All vital functions of an organism occur within cells. Because each individual cell is alive and carries on all of its own life processes, it is no surprise that these same vital functions carried out at the higher levels of organization are still done by the cells. As you move up the ladder of complexity, life processes are performed within and around the cells. But as more cells are added, more organization is required to meet their

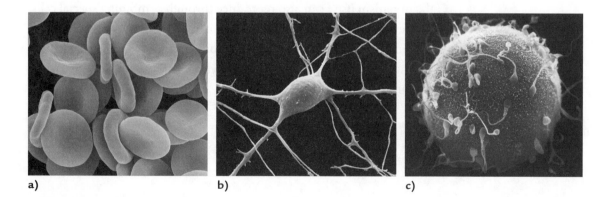

a)   b)   c)

FIGURE 6.2   **Human cells come in a variety of types and sizes. a)** This scanning electron micrograph clearly shows that red blood cells are biconcave disks. **b)** A nerve cell (neuron) has numerous cell extensions that can be quite long, as shown in this scanning electron micrograph. **c)** A human ovum (egg cell) is covered by very tiny sperm cells at fertilization.

collective demands. As organisms increase in size and complexity, cells become more specialized and join together to share functions, forming tissues and eventually organs. For example, your digestive system brings in adequate nutrients for all of your cells, and your circulatory system ensures that all of your cells are well-serviced and able to communicate with each other. While your cells perform their highly specialized functions in complex organ systems, your nervous system carefully choreographs them. All the specializations in structure and organization keep your individual cells alive and functioning in a highly efficient and synchronized manner, while the cells go about their business of life, maintaining not only themselves, but also you.

All cells come from preexisting cells. Most cells can reproduce. The normal cell cycle, which we will discuss, ends with the cell dividing to produce two daughter cells. Each of the daughter cells is almost identical to the parent cell and will quickly take on its function. So our body parts are constantly being replenished as old cells die. Similarly, cells divide to replace other cells that may be lost during an injury.

Cells contain hereditary information that regulates cell functions and is passed from generation to generation. Inside each of your cells is the blueprint for life—the DNA that houses your cells' genes. These genes determine what work your cells will do, and these instructions for life are passed on with reproduction. Not only do the cells divide, but the organism itself can also reproduce through combining special cells. Gametes—ova in females and sperm in males—unite at fertilization, yielding a new combination of DNA from both Mom and Dad, and producing a new offspring in which the whole process of life begins anew at many levels.

✔ **QUICK CHECK**

What are five principles that make up the Cell Theory?

_____

_____

_____

_____

_____

A Cell by Any Other Name . . . **Prokaryotes and Eukaryotes**

As mentioned earlier, cells can be quite diverse. Even within the human body we see tremendous variation in cell form, so imagine how much variety there is if we consider all living organisms! For example, plant cells have rigid cell walls and green chloroplasts, neither of which is found in humans. Despite this tremendous variation, all cells share at least three common characteristics:

1. They are enclosed in an outer **cell membrane,** which separates their internal environment from the external environment.

2. They are filled with **cytoplasm,** which is a mixture of substances in a liquid.

3. They contain deoxyribonucleic acid (**DNA**), which is the cell's genetic material.

Because cells come in many forms, there are many ways to classify them. One classification considers the basic organization of a cell's internal structure. All cells are either **prokaryotes** or **eukaryotes** (**Figure 6.3**). *Pro-* means before and *karyo-* means "*nucleus,*" so the term prokaryote means "before the nucleus." Prokaryotic cells lack internal membranes so, indeed, they don't have an enclosed nucleus. Instead, their DNA exists mostly in a loop and the area in which it tends to be located (but not enclosed) is referred to as the **nucleoid**. Additional DNA may be present in small loops, called **plasmids**, that can be transferred to other cells. Prokaryotes also lack most other internal cellular structures found in eukaryotes. Although these cells appear primitive, they remain all around us—all bacteria are prokaryotes.

Answer: All organisms are composed of one or more cells. Cells are the basic structural and functional units of life. All vital functions of an organism occur within cells. All cells come from preexisting cells. Cells contain the hereditary information needed to regulate cell functions and it is passed on to the next generation of cells.

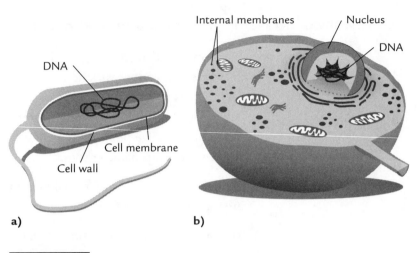

DNA

Internal membranes

Nucleus

DNA

Cell membrane

Cell wall

a)   b)

FIGURE 6.3 **Cells are classified as either prokaryotic or eukaryotic.**
a) Prokaryotes lack internal membranes and have no nucleus.
b) Eukaryotes have both.

WHY SHOULD I CARE?

Although the human body is composed entirely of eukaryotic cells, prokaryotes are also important. Some bacteria enter our bodies through the foods we eat and take up residence in our colons (large intestines), where they live off our dietary leftovers. While consuming these materials, the bacteria actually produce vitamins that we use. They also, unfortunately, release methane gas in our guts that can cause us to be uncomfortable or, if it escapes, perhaps a bit embarrassed! Other bacteria can cause infection and illness if they enter our systems.

The cells of the human body are all eukaryotic cells, or eukaryotes. The name comes from Latin: *eu-* means "true" and these cells have a true nucleus. The nucleus is a membrane-bound structure that houses the cell's DNA, keeping it localized. The cytoplasm in eukaryotes also has other structures with specialized jobs. Collectively these structures—the functional parts inside the cell—are called **organelles**. Some of these organelles are built from membranes; others are not. All eukaryotes contain some of the same organelles, but specialized cells often have unique features as well.

All human cells are eukaryotic cells. ■

✔ **QUICK CHECK**

In which type of cell would you find a nucleoid, and how is it different from what would be present in the other type of cell?

_____

_____

_____

Paper or Plastic, Ma'am? **The Cell Membrane**

Your cells have certain requirements that must be met for them to give their peak performance. For example, they need the right amount of fluid, nutrients, water, and oxygen. Recall that **homeostasis** means maintaining a relatively constant internal environment (which you can think of as optimal working conditions). Doing so is not always easy.

One feature that helps achieve the goal of homeostasis is the cell membrane, also called the **plasma membrane**. It is simultaneously a container that holds a cell together and a physical partition that separates the cell's inside world from the outside, making it easier to maintain constancy inside. The heading of this section, "Paper or Plastic," conjures up the image of a grocery sack. That image works for the concept of the cell membrane as a physical separation between the cell's inside and outside worlds. Clearly, this separation makes it easier to maintain homeostasis inside the cell even if conditions outside the cell vary.

Answer: A nucleoid is found only in prokaryotic cells and it is where the DNA loop is located. Eukaryotic cells, instead, have a membrane-bound nucleus that encloses their DNA.

But if the membrane was truly like an actual sack and formed a complete barrier, nutrients could not enter your cells, nor could wastes and products manufactured by your cells leave. Thus, the cell membrane must have unique characteristics that allow some materials to pass through while blocking others.

The cell membrane is made primarily of **phospholipid** molecules. These molecules have phosphorous and other atoms at one end, forming what is called the **phosphate head**. This portion of the molecule is polar, or **hydrophilic**. Look back to the word root and suffix tables in Chapter 3. What does hydrophilic mean? _____

_____

Attached to one side of the phosphate head are two longer molecules called **fatty acid tails**. These tails are the main reason that the molecule is a lipid, and you should recall from the last chapter that lipids are non-polar, or **hydrophobic**. What does that mean? _____

_____

To envision how a phospholipid molecule is organized, think about a brass brad—an old-fashioned paper fastener you likely used in school when you were young or that may even be built into some of your folders now (**Figure 6.4**).

TIME TO TRY

Let's figure out how phospholipid molecules arrange in a cell membrane.

1. What happens when a lipid comes into contact with water? Fill a clear container about 2/3 full of water. Watching closely, add 1/2 teaspoon of cooking oil. Describe what happens when the oil first enters the water, and where it ends._____

_____

2. Cover half of a slice of bread with a thick layer of peanut butter, which has a very high fat (lipid) content. Gently place a drop of water on the bare bread. What happens? _____

_____

Place a drop of water on the peanut butter. What happens? ____

_____

How do these observations relate to the organization of phospholipids in a cell membrane? _____

_____

Realize that a cell has an inside and an outside, and both contain water. In humans, the space outside of your cells contains extracellular fluid that is mostly water. The cytoplasm inside your cells also contains large amounts of water. The phosphate heads of the phospholipids interact fine with water, but the opposite ends of the molecules—the fatty acid tails—do not. As you saw when you placed oil in water, lipids in contact with water first form a ball, then settle into a single layer at the surface with one side in contact with the air, not water. In a cell, though, both surfaces are in contact with water. How can the phospholipid molecules arrange themselves so the fatty acid tails don't contact water?

_____

Think about that partial peanut butter sandwich you made earlier. You should have seen the water readily enter the bare part of the bread, but ball up on the peanut butter. How could you organize bread and peanut butter in such a way that there are two major surfaces, or sides, and water will not be repelled from either one? _____

_____

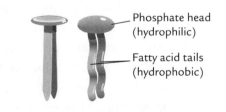

Phosphate head
(hydrophilic)

Fatty acid tails
(hydrophobic)

FIGURE 6.4 **The cell membrane is a phospholipid bilayer.** To understand the organization of a phospholipid molecule, think of a brass brad that is used to hold paper together. There is a head with two long parts hanging off of it. The head represents the phosphate "head" of our molecule, which is hydrophilic. The prongs that hang down represent the lipid "tails," which are hydrophobic.

I hope you guessed it—you have likely eaten it plenty of times. Consider the basic peanut butter sandwich. Assume you made it with two slices of bread, each smeared with a thick layer of peanut butter, then sandwiched together. You would have bread on both sides of the sandwich and peanut butter in the middle—two layers of it facing each other. This is the basic structure of the **phospholipid bilayer** (*bilayer* = two layers). The hydrophilic phosphate heads (bread) are arranged in two layers so that they face the water in the extracellular fluid and in the cytoplasm. The hydrophobic tails (peanut butter) are sandwiched in the middle, out of contact with the water. This is how the cell membrane is organized (**Figure 6.5**).

But cell membranes don't contain only phospholipids—there are other molecules as well. Cell membranes contain cholesterol molecules, for example, which help stabilize the membrane. Some carbohydrate molecules act as labels that allow your cells to recognize other cells, such as when sperm are trying to locate an ovum for fertilization or when your immune system is destroying foreign invaders to keep you healthy. These are but a few examples.

Cell membranes also contain assorted proteins. Some of these proteins form channels that act like tunnels that allow certain molecules to pass through the membrane. Other proteins might be carriers that pull various molecules through the membrane. The combination of phospholipid molecules and specialized protein channels and carriers determines what can and cannot pass through the cell membrane. Because not everything can pass through, the cell membrane is said to be **selectively permeable**—its chemical composition restricts the movement of some substances, so it is selective about what can pass through.

All of these assorted molecules are positioned throughout the cell membrane in what can be thought of as a sea of phospholipid molecules (see Figure 6.5). Imagine a child's pool filled with water, and floating in it are toy boats, inflatable toys, and a few children. The objects are free to move around in the water, and in fact they do. This is the nature of the cell membrane. This model, known as the **fluid mosaic model**, reveals a cell membrane that is very dynamic, moving, changing, and fluid in nature.

The cell membrane is composed of a phospholipid bilayer but also contains other molecules, and it is a very active structure. ∎

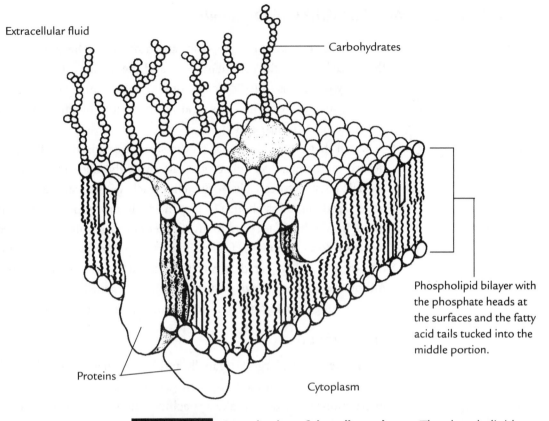

Extracellular fluid

Carbohydrates

Phospholipid bilayer with
the phosphate heads at
the surfaces and the fatty
acid tails tucked into the
middle portion.

Proteins

Cytoplasm

FIGURE 6.5 **Organization of the cell membrane.** The phospholipid
molecules are organized into a bilayer in which other molecules are
embedded.

JUST FOR FUN

Borrow a child's bottle of bubbles or, if no child is available, fashion
your own wand by making a loop out of a large paperclip or a
length of wire—those twisty ties for garbage bags work well. Make
your own bubble solution by mixing some liquid dishwashing soap
in a bowl or cup of water. Now, blow! Be sure there is plenty of
overhead light as you observe the bubbles. Notice how the colors on
the wall of a bubble constantly swirl around and change. This is the
way the cell membrane is—constantly in motion and changing. See
why it is wrong to think of it as just a sack?

What Department Are You With? **Cell Organelles**

Eukaryotic cells contain cytoplasm and a nucleus. The cytoplasm fills the space between the cell membrane and the nucleus. The liquid part of the cytoplasm is called **cytosol,** and it contains many dissolved substances, such as nutrients. It is a rather thick fluid in which numerous cell structures are suspended.

Recall that a typical cell has all the structures and does all the work needed to maintain life. Eukaryotic cells divide the labor among different specialized structures called organelles, each with a particular job. To understand this, think about a factory in which some product is manufactured. Inside this factory are many different departments, all involved in some way to make sure the factory is working properly and that the product is made and delivered. In a cell, the departments are the organelles. Although each has its own particular task, all organelles are coordinated and work together. **Figure 6.6** illustrates the main structures of a cell.

THE NUCLEUS

The **nucleus** is the largest organelle and it really has one specific task: it houses the DNA that contains your genes. Each gene is essentially the instructions for how to make a specific protein. All cells in your body contain two copies of each of your genes—one from Mom and one from Dad. The only exceptions to this are your sex cells—ova or sperm—which contain only one copy of each of your genes. This copy will combine with a second copy from your partner if conception occurs. Your DNA is organized into threadlike strands, called **chromatin,** that condense into rodlike structures called **chromosomes** when the cell reproduces (more about this later).

Your DNA determines what proteins can be made in your body, but only certain proteins are made by each of your cells. For example, the cells on your elbow don't usually grow moustaches, but those on your chin might! The genes are activated and deactivated within the nucleus based on many different factors, and the activation determines the cell's productivity. Thus, in your cell "factory," the office is the nucleus where the boss—your DNA—determines what needs to be done.

The nucleus is enclosed in a **nuclear envelope** made of a double membrane, each of which is similar to the cell membrane. This envelope is

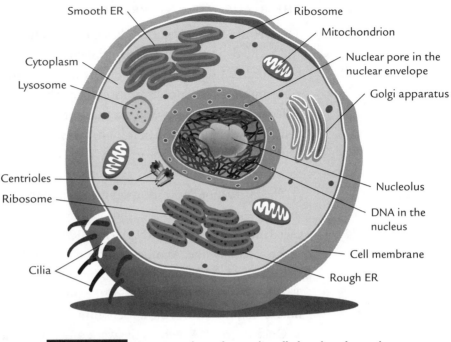

FIGURE 6.6 **A composite eukaryotic cell showing the major organelles.**

pierced periodically by holes called **nuclear pores** that make it more permeable than the cell membrane, allowing larger molecules to pass through. This is important because the instructions—the genes—never leave the nucleus. To get the daily work assignment, a memo is sent out to the workers—the boss does not come to them. Inside the nuclear membrane is a jellylike liquid called the **nucleoplasm**, which is similar to the cytoplasm.

✔ **QUICK CHECK**

What is the main function of the nucleus? _____

_____

 DNA is your genetic material and it determines what proteins your cells can make. ▪

Answer: The nucleus houses the DNA.

THE RIBOSOME AND PROTEIN SYNTHESIS

The DNA in the cell's nucleus has the instructions for all proteins that will be made in the cell, but the proteins are not manufactured in the boss's office. A separate cellular department—a tiny organelle called the **ribosome**—is responsible for assembling the protein during a process called **protein synthesis.** You can think of a ribosome as being the work bench on which the protein is built, rather similar to an assembly line. The workers who build the proteins there are called **RNA**, which stands for **ribonucleic acid.**

The instructions or work assignments from the boss—the DNA in the nucleus—are carried to the ribosome by a special worker, a molecule called **messenger RNA (mRNA).** The mRNA carries the work assignment *to* the assembly line. Another worker, a type of RNA called **ribosomal RNA (rRNA)**, is always *at* the assembly line—it is part of the ribosome. This RNA is actually made inside the nucleus at a structure called the **nucleolus,** and it leaves the nucleus through those large nuclear pores.

Once the mRNA delivers the work order from the DNA to the ribosome, the specific proteins are made at the ribosome by linking together small molecules called amino acids, which you should recall from Chapter 5. The amino acids are carried to the ribosomal assembly line by other workers called **transfer RNA (tRNA),** then assembled according to the DNA's instructions. RNA's job is to build the protein the DNA tells it to make. One type of RNA carries DNA's instructions to the ribosome, and the other two types of RNA are directly involved in building the protein according to those specifications.

THE ENDOPLASMIC RETICULUM

Some ribosomes are free to float around the cytoplasm; others are attached to yet another organelle called the **endoplasmic reticulum,** or **ER.** The ER is an extensive network of membranous tubes and channels inside the cell. Ribosomes look like tiny dots, so when they are attached to the ER they give it a rough appearance; thus it is called rough ER. The presence of ribosomes tells you that at least part of rough ER's job is protein synthesis. Smooth ER lacks ribosomes. Instead of making proteins, it is involved in making other materials (like lipids), detoxifying potentially harmful substances, and transporting materials around the cell.

You can think of the ER as being like a system of workers who are always passing through various hallways, moving materials from one work station to another, and throughout the factory. Because the ER connects different parts of the cell, it also provides a communication network within the cell. Materials are brought into this system, moved around, changed, and turned into raw, rough product here.

GOLGI APPARATUS

The **Golgi apparatus** looks like a stack of flattened membranous sacs. This is the processing, packaging, and shipping department in our cells. Products that have been made elsewhere in the cell, such as at the ribosomes or in the ER, are sent here to be finished. They are modified and put into their final forms. They are slapped with a molecular "shipping label"—a chemical tag that determines where they will go. They are packaged in some membrane that pinches off of the Golgi and surrounds the product, forming a saclike structure called a **vesicle**. Finally the products are shipped, some to other parts of the cell, some to the cell membrane, and some out to the great extracellular world beyond.

✔ **QUICK CHECK**

1. Where in a eukaryotic cell would you find ribosomes? _____

_____

2. What is the functional relationship between the nucleus, nucleolus, ribosome, rough endopalsmic reticulum, and Golgi apparatus?

_____

_____

_____

Answers: 1. Free-floating in the cytoplasm or attached to endoplasmic reticulum. 2. The nucleus houses DNA that has instructions for how to build proteins. Proteins are built at the ribosome, part of which is made in the nucleolus inside the nucleus. Many ribosomes are located on the rough ER. Proteins made at the ribosomes move into the Golgi apparatus to be processed, packaged, and shipped to their final destinations.

THE MITOCHONDRION

If your factory is to do its work 24/7, it needs a good power source. The **mitochondrion** is your cell's powerhouse. It provides a constant supply of energy to drive the work being done throughout the cell. As discussed in Chapter 5, cellular energy comes from the foods we eat, where it is stored in the chemical bonds that hold the food's atoms together. Specialized chemical reactions in the mitochondria harness that energy and store it in a molecule called **ATP**—*a*denosine *tri*phosphate (see why we call it ATP?). All cells, in any organism, use ATP directly for energy. Think of it as the electricity that powers our cells. The more work a cell is doing, the more ATP it needs, and the more mitochondria it will have.

Mitochondria are unique organelles (**Figure 6.7**). They contain their own genetic information and can reproduce. These elongated organelles are enclosed by a double membrane, similar to the nuclear membrane. The outer membrane is smooth, but the inner membrane is highly folded. In general, such folding occurs to increase the surface area of the membrane, and increasing the surface area is the same as increasing the workspace. It is rather like expanding a department.

TIME TO TRY

To understand how the foldings of the mitochondrial membrane increase the surface area, get a small plastic bag. This represents the outside membrane of the mitochondrion. Notice that it is smooth all the way around. Now, get a larger plastic grocery bag and note its size. This represents the inner folded membrane of a mitochondrion. Can you fold it enough to fit inside of the original bag? If you fold it enough, it should fit. Now realize how much more membrane can be used if it is highly folded.

LYSOSOMES

While the organelles are doing all this hard work in your cellular factories, they need some assistance in keeping their workplace tidy. **Lysosomes** are your cell's janitorial staff. A lysosome is a small membranous bag containing strong digestive enzymes. Its main job is to

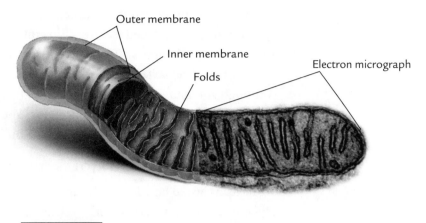

Outer membrane

Inner membrane

Electron micrograph

Folds

FIGURE 6.7   **A mitochondrion.** This illustration includes a drawing (left) blended into an actual electron micrograph (right).

break down materials. Some materials are brought into your cell and digested to provide basic building materials for your cell to use in its work. Other materials may be old, worn out cell parts or foreign material that invades your cells. Lysosomes destroy these items so they neither harm nor clutter your cell's interior. A lysosome's job is to recycle what it can and get rid of the remaining garbage.

PEROXISOMES

If lysosomes are your cell's janitorial staff, **peroxisomes** are its hazardous waste disposal team. Peroxisomes break down organic compounds, such as fatty acids, whose breakdown produces hydrogen peroxide (H_2O_2). You're probably familiar with the bubbling that H_2O_2 causes when you use it to disinfect a wound. That demonstrates its high reactivity—it could do a lot of damage to your cells. Instead, the peroxisome conducts its work inside its own membrane and, conveniently, contains enzymes that break down the H_2O_2, rendering it and other chemicals harmless to your cell. How's that for service?

THE CYTOSKELETON

Remember that all of this work is going on inside your cells, in a liquid environment. The "building" needs a frame to hold it up. We discussed the outer partition—the cell membrane. But we need something inside to hold the membrane out so the cell does not collapse on itself. The **cytoskeleton** is composed mostly of tiny tubes (**microtubules**) and filaments (**microfilaments**). These structures form a type of scaffolding that supports the cell and to which various organelles are attached. Although the name sounds like this structure is made of bone, the cytoskeleton is actually made of proteins. You can think of them as being the struts and beams that hold up the building, or cell.

CELL MOVEMENT: CENTRIOLES, CILIA, AND FLAGELLA

We have discussed how materials can move through a cell. But there are other movements associated with a cell, too. An area of the cell called the **centrosome** ("central body") is composed of paired cylindrical structures made of microtubules. These structures are called **centrioles**. They direct the movement of the chromosomes when a cell reproduces, as we will discuss shortly, but they also form part of two other structures involved in cell movement: cilia and flagella.

Cilia look like fringe on a cell. Not all cells have cilia, but the ones that do have several. They are extensions of the cell and they are mobile. Cilia are coordinated so that they tend to move in a wavelike manner. Cilia sweep materials over the outer surface of a cell, moving materials past the cell. For example, cilia in your respiratory tract help clear debris so it doesn't clog the air sacs where oxygen enters your blood.

WHY SHOULD I CARE?

Cigarette smoke paralyzes the respiratory cilia for about an hour per cigarette. During that time, they cannot prevent the particulate material that we breathe in from reaching the lungs. Over time, more of this material clogs the small air sacs and begins to damage them. People who experience this must cough, especially upon arising in the morning, to try to clear the material that has accumulated in their lungs. This is the basis of *smoker's cough*.

PICTURE THIS

Imagine you are at the "big game" and the crowd is tossing a beach ball around. As this is going on, off in the distance you see the crowd start a "wave," where they stand and wave their arms overhead, then sit, group by group, all around the stadium. The wave approaches you just as the beach ball is heading your way. The wave passes you and so does the beach ball. It was carried away on the wave and you see it now making its way around the stadium, riding the wave. This is how cilia move materials across the surface of the cell—they beat in a coordinated manner, sweeping materials along.

In contrast to the cilia, a **flagellum** is a single, long, tail-like extension of the cell. In humans, these are found only on sperm cells. A flagellum whips back and forth to propel the sperm through the male reproductive tract and up into the female's tract in search of an ovum to fertilize.

We have reviewed the organelles of a typical eukaryotic cell, like our human cells. You will learn more detailed descriptions of these structures and their jobs as you proceed in your A & P course. For now, you should have a general understanding that will give you a knowledge base from which to work. **Table 6.1** summarizes the organelles we have discussed and provides a quick review.

✔ **QUICK CHECK**

1. Flagella use tremendous amounts of energy to propel the cell forward. Which organelle would you expect to see in large numbers near the very busy flagellum? _____

2. How do the movements of cilia and flagella differ? _____

_____

_____

Answers: 1. Mitochondria. 2. Cilia are numerous and they beat in a wavelike manner to sweep materials across the cell surface; a flagellum is a single whiplike process that propels the entire cell.

TABLE 6.1 **Summary of cell organelles and structures.**

Organelle	Description	Function
Nucleus	Rounded larger membranous sac with pores.	Houses the DNA that directs cellular activities.
Chromatin	Relaxed strands of DNA in the nucleus.	Contain genes that determine what proteins can be made in the cell.
Ribosome	Small nonmembranous structure free in the cytoplasm or attached to ER.	Protein synthesis.
Nucleolus	Small body located in the nucleus.	Makes part of a ribosome.
Endoplasmic reticulum (ER)	Extensive network of membranous tubes and channels.	Protein synthesis (rough ER); lipid synthesis (smooth ER); detoxification; communication and transport system.
Golgi apparatus	Flattened stack of membranous sacs.	Processing, packaging, and shipping of cellular products.
Mitochondrion	Elongated membranous structure with highly folded internal membrane.	Powerhouse; harnesses energy from food molecules and stores it in ATP.
Lysosome	Small membranous sac.	Breakdown of unwanted materials; recycling of molecules.
Peroxisome	Small membranous sac.	Breaks down organic compounds and neutralizes H_2O_2 created in the process.
Cytoskeleton	Meshwork of microtubules and microfilaments.	Provides support and structure to the cell interior; anchors organelles.
Centrioles	Cylindrical structures made of microtubules.	Direct movement of chromosomes during cell reproduction; form part of cilia and flagella.
Cilia	Small, numerous, hairlike processes that beat in a wave.	Sweep materials over the surface of the cell membrane.
Flagellum	A single whiplike tail.	Propels the cell forward.
Vesicle	Small membranous sac.	Contains materials entering or leaving a cell.

Tunnels and Doorways: **Movement Processes**

Atoms and molecules are always moving, both inside and outside the cell. Cell products packaged at the Golgi apparatus are shipped out of the cell to travel elsewhere in the body. Nutrients and building blocks are moved in from the outside. Waste products must leave. Let's examine how some of these movements occur.

BROWNIAN MOTION

Atoms and molecules constantly move in a random manner called **Brownian motion**. This is a rather nondirectional, jiggling movement. Remember those bumper cars we had fun with before? If you're in a car and someone hits you, their car bounces off in a new direction, while your car careens off in yet another. This is similar to how atoms and molecules move. Because they are constantly in motion, you can imagine how particles bump into each other and ricochet away.

CONCENTRATION GRADIENTS AND EQUILIBRIUM

Although Brownian motion is random, other movement processes are not. Before we discuss these movements, though, we need to understand the concepts of concentration gradients and equilibrium. You can think of concentration as referring to how crowded together molecules are—the more crowded they are, the more concentrated they are. A **concentration gradient** exists whenever there is a difference between the concentrations of the molecules in two areas (**Figure 6.8a**). Perhaps your vehicle has gradient glass in its windshield—it is tinted with more color at the top than at the bottom, which means there is a *gradient* in the color.

If, instead, there is about equal space between all of the molecules, we say they are at **equilibrium** (**Figure 6.8b**). Think about standing in an elevator. If there are only two of you on the elevator, you likely stand at opposite sides. Add two more people and you likely stand one in each corner with about equal distance between yourselves. As you add more and more people to the elevator, you become more closely crowded and are more likely to bump into each other. And if you do collide, you respond by moving away, always trying to maximize your personal space. That is how atoms and molecules move, but not by conscious decision. They move by basic laws of science. Let's explore how.

SIMPLE DIFFUSION

One basic type of molecular movement is **simple diffusion**. This is how oxygen enters your cells and carbon dioxide leaves them, for example, and it is critical for sustaining life. Simple diffusion occurs when molecules move from an area of higher concentration to an area of lower concentration. In other words, they move from where they are more crowded to where they have more room. Think of it simply as the molecules spreading out. Whenever there is a concentration gradient, molecules will spontaneously move *down the concentration gradient*—that means they move from where they are most concentrated, or crowded, to where they are least crowded. You are already familiar with diffusion. Consider, for example, baking brownies. The molecules that produce the yummy smell quickly diffuse through the air in your house so that any visitors know you have a treat to share!

With individual molecules the motion is random but, like people in an elevator, the more molecules that are present in an area, the more likely it is that they will bump into each other and bounce away. There will be more collisions in an area where the molecules are more crowded, sending them skidding off. They will ricochet less as they move into areas where there are fewer molecules to bounce off of. In time the molecules, like the people on the elevator, will have moved

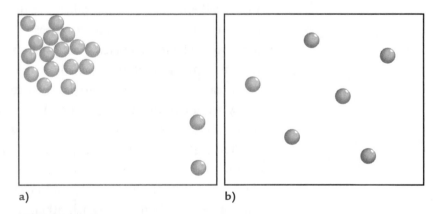

a)   b)

FIGURE 6.8   **a)** A concentration gradient exists when there is a difference in the concentration, or spacing, around molecules in two different areas. **b)** Equilibrium exists when the molecules are spaced about evenly.

around enough to have almost equal distance between them. This is equilibrium. However, unlike the people on the elevator carefully maintaining their positions, molecules at equilibrium do not stop. They continue moving, but all molecules experience about the same number of collisions and they maintain a fairly even spacing.

Molecules diffuse at different rates under different conditions. Molecules diffuse faster when there is a greater concentration gradient between the two areas. Molecules in high concentrations diffuse faster than those in lower concentrations. Smaller molecules also move faster than larger molecules, and temperature alters the diffusion rate as well.

TIME TO TRY

Let's see how temperature affects diffusion. Get three clear glass containers of about the same size. Fill one halfway with very hot water, another with room temperature water, and the third with very cold water. Wait until the water stops moving, then gently add a drop of food coloring to each container. Or you could instead use a tea bag. Now just observe. You should see evidence that the molecules are diffusing—the color should spread out from where it is most concentrated. Eventually you should see the equilibrium state—all of the water should be of uniform color, meaning that the molecules have spread out equally, even though they continue to move.

How did the diffusion rates differ, and why? _____

_____

You should see that increased temperature also increases the diffusion rate. This is because heat makes molecules move faster.

FACILITATED DIFFUSION

Simple diffusion is only one mechanism that allows materials to enter and leave a cell. Only lipid-soluble nonpolar molecules can diffuse directly through the cell membrane. Larger polar molecules, such as glucose (one type of sugar) cannot diffuse through the membrane as easily. Instead, they are moved by a special protein carrier molecule in the cell membrane. The molecules still move from an area of high concentration to one of low concentration, trying to reach equilibrium. The carrier molecule merely helps, or facilitates, the molecules' movements through the cell membrane. It acts like a special door through which they can pass. This type of diffusion is called **facilitated diffusion**.

 Diffusion is a passive process by which molecules spontaneously move from where they are in high concentration to where they are in low concentration. ▪

OSMOSIS

Osmosis is a special type of diffusion that is also critical to our survival. It is the diffusion of water through a selectively permeable membrane, such as the cell membrane. We've discussed the lipid nature of the cell membrane, so it seems surprising to discover that water actually passes through it fairly well. This is partly due to the fluid nature of the molecules in the membrane, but for our purposes the specifics aren't critical. Before proceeding, though, let's recall the basics of solutions.

TIME TO TRY

Let's head to the kitchen. Get a clear drinking glass or measuring cup and fill it about 2/3 full with warm water.
Describe what you see. (Humor me, OK?) _____
Next, get a spoonful of sugar. Describe what you see. _____

Finally, add the sugar to the water and mix it thoroughly. What do you see? _____

This exercise may have seemed silly, but you just made a solution by combining a solute and a solvent. The sugar is the **solute**—the material that gets dissolved. As in our bodies, water is the **solvent**—the material that dissolves the solute. The end result is a **solution**, which is a homogeneous mixture—it should look uniformly clear throughout, because you cannot see the dissolved solute.

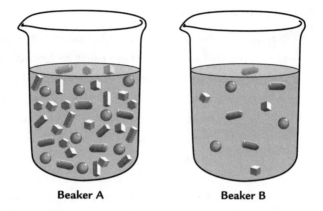

Beaker A   Beaker B

FIGURE 6.9 A solution contains both a solvent, such as water, and solute molecules. In this illustration, the beakers contain the same amount of solution. The geometric shapes in each solution represent various solute molecules. The remaining shaded space represents the water. Clearly, beaker A, which contains more solute molecules, has less space left for water, but beaker B, with fewer solute molecules, contains more water.

To understand the relationship between water and solute molecules, recall that matter is anything that has mass and occupies space, so no two molecules can occupy the same space at the same time. Let's think about that sugar solution you just made. If you started by first putting a half-cup of sugar in the container, there would have been less room left to fill with water. The opposite would also be true—less sugar would allow more water. And it doesn't matter what the solute molecules are—they each take up their own space (**Figure 6.9**). That is why I have adopted, in my own classes, a highly technical term for the various solutes: *stuff*. When we are talking about osmosis through a cell membrane, we must look at the water and the other stuff for a simple reason—the more stuff you have, the less water there can be because the stuff takes up space that water can no longer occupy.

TIME TO TRY

Prove to yourself that when there are more solute molecules in a given space, there must be less solvent, and vice versa. In other words, the more concentrated the solute, the less concentrated the solvent (water). Get a glass measuring cup, then grab a handful of household items—paper clips, coins, marbles, and so on. The only rules are that the objects cannot float and they cannot dissolve (most food items are not good for this reason). We had better add one more rule—they should be waterproof! These objects represent the solute molecules—the stuff—and the water, of course, is the solvent. Drop all of the objects into the cup, then fill it with water to the one-cup mark. Look at the layers of water and objects and note the sizes of each layer. Carefully pour the water into a second container and put it aside. Place the objects on some paper towels, then pour the water back into the cup to measure it. How much water was in the solution? _____

Repeat this procedure, but this time place only about half of the items in the cup before filling it with water to the one-cup mark. Again, pour the water into your second container, remove the objects, then pour the water back into the cup to measure it. How much water did your mock "solution" contain this time?

_____

In which trial did you use more "solute molecules" (objects)?

_____

In which trial did you have the least water? _____

Explain the relationship between the amount of solute molecules and the amount of solvent present in a solution. _____

_____

_____

Recall that molecules never stop moving, even after reaching equilibrium. Each of your cells has two solutions separated by a selectively permeable membrane—the extracellular fluid outside of the cell and the

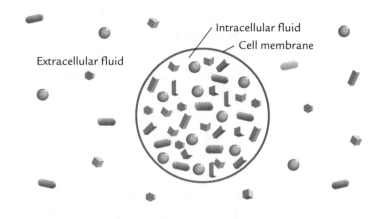

FIGURE 6.10 **The cell membrane separates the extracellular fluid from the intracellular fluid.** In this figure, a situation is shown in which there are more solute molecules inside the cell than there are outside of it. Assume the shaded shapes are solute molecules and the remaining space is filled with water.

intracellular fluid inside the cell. These solutions have different compositions. If your cell membranes were fully permeable, all molecules could pass through and move down their concentration gradients until reaching equilibrium, and we would have simple diffusion. But the cell membrane is semipermeable. We know water can pass through the cell membrane rather easily, but many solute molecules can't because of their size or their chemical composition. You also know from your previous activity that there is a higher concentration of water (solvent) wherever there is a lower concentration of other stuff (solute molecules).

Look at **Figure 6.10**. There is a higher concentration of solute molecules inside the cell than there is outside. Assume the solute molecules can't pass through the cell membrane. Now answer these questions:

1. Which fluid contains the lowest concentration of solute molecules?

   _____

2. Which fluid has the highest concentration of water? (Recall that when water is high, solute is low, and vice versa.)

   _____

By osmosis, water diffuses—it moves down its concentration gradient—from where it is in higher concentration to where it is in lower concentration. You should see clearly from Figure 6.10 that, in this example, there is a greater concentration of water in the extracellular fluid than there is in the intracellular fluid, so water will enter the cell.

For osmosis, you just need to know where the highest concentration of water is, because it will always move from that area to the area with less concentration. The only reason to consider the other stuff (solute) is because it tells you where the water is—the area with the lower solute concentration has the higher water concentration, so the water will move away from that area.

Osmosis is diffusion of water through a selectively permeable membrane. ■

What happens to cells when water moves across the cell membrane? **Figure 6.11** shows three views of red blood cells. Figure 6.11a shows a normal red blood cell, shaped like a biconcave disk. If water leaves a cell, the cell will shrink and the cell won't function efficiently. This is shown in Figure 6.11b. If water moves into a cell, the cell will swell. This is shown in Figure 6.11c. This swelling increases the pressure inside the cell and impedes normal function.

PICTURE THIS

You are already familiar with water moving into and out of cells and changing their shapes and sizes. Unlike our cells, plant cells are surrounded by a rigid cell wall that helps maintain their shape. Have you ever had a houseplant or a plant in your yard that was a bit wilted and droopy? When the plant cells lose water, they shrink and cannot support the weight of the plant parts above them, so they droop. When they receive water, the cells enlarge, exerting pressure against the rigid cell wall, and this pressure helps support the rest of the plant—it stands back up.

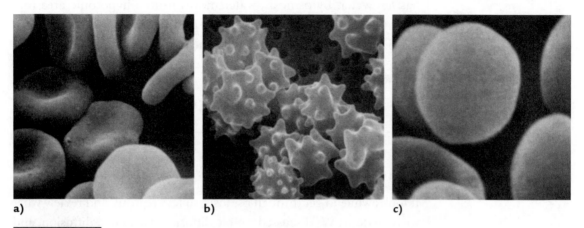

a)   b)   c)

FIGURE 6.11   **Red blood cells showing the effects of osmosis. a)** Normal red blood cells. **b)** Red blood cells that have lost water through osmosis. **c)** Red blood cells swollen from taking in too much water by osmosis, and at risk of rupturing.

WHY SHOULD I CARE?

Animal cells lack a cell wall, so they are more likely than plant cells to rupture if they take in too much water. In health care, IV (intravenous) fluids must have the appropriate amount of water—if there is more water in the IV fluid than in the patient's intracellular fluid, his or her cells will swell. If there is less water in the IV fluid, the patient's cells will shrink. Unless the patient is being treated for a fluid imbalance, IV fluids should have the same water concentration as intracellular fluids so the water will be at equilibrium. Such fluids are called *isotonic*. The wrong amount of water in an IV fluid can be fatal.

When comparing solutions, we often look at their concentrations, meaning their solute content. Two solutions with the same concentration are said to be **isotonic** to each other. But if their concentrations differ, the one whose solute concentration is highest is hypertonic, and the one with the lower solute concentration is hypotonic. (Recall from Chapter 3 that *hyper* means *over* or *above*, and *hypo*, then, means *under* or *below*.) Beware, though—when considering osmosis, you must focus on the water. A hypertonic solution has more solute, but that means it has *less* water. By osmosis, water moves from a hypotonic area to a hypertonic area. Perhaps it is easier to think of it this way—hypotonic solutions are weaker or more dilute, meaning they have more . . . *water!*

✔ QUICK CHECK

Refer back to Figure 6.10. Will water enter or leave the cell?

ACTIVE TRANSPORT

It took awhile to get through osmosis, so let's regroup and review what we have done. We discussed simple diffusion, facilitated diffusion, and osmosis. These are all types of diffusion, so the molecules being examined will move from an area where they are more concentrated to an area where they are less concentrated. Diffusion is a spontaneous process—it happens automatically. These types of diffusion do not require energy, so they are referred to as **passive transport**.

Sometimes, instead, cells need to move molecules against their concentration gradients. For example, for a nerve cell to transmit a signal, sodium ions must enter the cell. Special channels in the cell membrane open and sodium ions diffuse into the cell because there are more sodium ions outside than inside. This is simple diffusion. After that nerve signal is sent, though, the sodium ions must leave so the cell is ready for the next signal. But there are still more sodium ions outside than there are inside. The ones inside can't merely diffuse out—they would be moving up their concentration gradient.

PICTURE THIS

Assume you are a dedicated anatomy and physiology student and you've been studying much more than you've been cleaning house. You hear a van pull up outside and you see that it is the *Prize Patrol* from a major sweepstakes sponsor, with a camera crew. They are

Answer: The water moves from where it is more concentrated to where it is less concentrated, so it moves *into* the cell.

broadcasting live and they are quickly approaching your door. You grab an armfull of clutter and all your books and open your closet door to stash them on the overhead shelf, but the shelf is full because it has already been jammed with previous clutter.

1. What likely happens if you just open the closet door and stand back? _____

2. Is it easy or difficult to cram more clutter onto the closet shelf?

   _____

3. Does the clutter spontaneously head onto the shelf on its own, or do you have to really work at getting it to go and stay there?

   _____

In the situation with the closet, when you open the door there is more clutter inside than there is outside, so the clutter inside will spontaneously fall out. This represents diffusion—molecules moving from where they are more concentrated to where they are less concentrated. You likely have to use your arm or somehow exert effort to prevent the clutter from falling out. At the same time, you have to force the additional clutter in your arms into the closet, and you have to work hard to do it. Work requires energy, so you must use energy to do it.

With cells, molecules will not spontaneously move against their concentration gradients. Work must be done, so energy must be used. For this reason, this type of movement—moving molecules up, or against, their concentration gradient—is called **active transport**. This type of movement also requires special one-way "doorways" in the cell membrane, called **pumps**, that ensure the molecules can only move in one direction. Otherwise molecules on the opposite side would spontaneously diffuse out, which is not our goal.

 Active transport moves molecules against their concentration gradient, which requires a special molecular pump and energy. ▪

EXOCYTOSIS

We have discussed ways for moving molecules, but now it is time to think bigger! Recall that cells may make proteins and other molecules that will be exported. These products are usually wrapped in a membranous sac, called a **vesicle**, at the Golgi apparatus. The method by which they are expelled is called **exocytosis** (*exo-* = outside, *cyto-* = cell). The method is rather simple. The vesicle makes its way to the edge of the cell and its membrane fuses with the cell membrane. As the vesicle pushes against the cell membrane, its own membrane ruptures and seems to peel back, becoming part of the cell membrane and releasing its contents. **Figure 6.12** shows a cell secreting a product via exocytosis. Notice the contents spewing out of the cell.

ENDOCYTOSIS

Endocytosis (*endo-* = inside) is the reverse of exocytosis. It is a way for cells to take in larger objects or even liquid that contains dissolved materials, such as nutrients. There are three major types of endocytosis:

- phagocytosis,

- receptor-mediated endocytosis, and

- pinocytosis.

Phagocytosis is the process by which solids are moved into your cells, and is sometimes referred to as "cell eating." This process demonstrates the active nature of the cell membrane. Extensions of the cell membrane, called **pseudopodia** ("false feet"), seem to reach out from the cell surface, rather like tiny arms, on each side of the object to be taken in. Then the extensions fuse and form a membranous sac, like a vesicle, around the object. This saclike structure is called a **phagosome**, and it moves inward, pinching off from the cell membrane on the inside of the cell. The object is now inside your cell. Soon, several lysosomes typically fuse with the phagosome. Some cells in your immune system are specialized for phagocytosis and use this process to rid your body of foreign material, such as bacteria, that might make you ill.

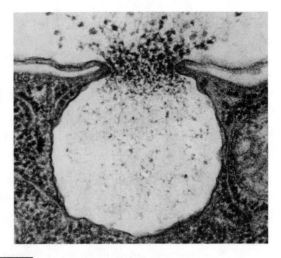

✔ **FIGURE 6.12** **A cell performing exocytosis.**

QUICK CHECK

Why do lysosomes fuse with the phagosome? _____

_____

In **receptor-mediated endocytosis,** before the cell can take the materials in, the objects to be moved must bind to special receptors in your cell membrane. This binding triggers the cell membrane to form a pocket that will enclose the materials and bring them into your cell. From this point on, the process is very much like phagocytosis.

Pinocytosis is a way your cells can bring in liquids and is sometimes referred to as "cell drinking." In this process, part of your cell membrane puckers inward, forming a pouch at the surface that contains extracellular fluid. In that fluid are a variety of dissolved substances that are now surrounded by cell membrane, which pinches off to form a vesicle-like structure called an **endosome.** The contents are then released inside the cell and are available for its use.

Answer: Lysosomes contain enzymes that will break down the contents of the phagosome, rendering potential threats harmless and recycling materials that are then made available to the cell for reuse.

Ashes to Ashes, Cells to Cells: **The Cell Cycle**

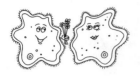

We have discussed many processes that occur in a cell, but a very important one remains: the cell life cycle. Remember from the Cell Theory that cells come from preexisting cells. Each cell goes through the **cell cycle** (**Figure 6.13**), and the duration of this cycle varies with the cell type. The cell cycle is divided into multiple phases. Simply, though, it can be viewed as having two main parts: **interphase** and **cell reproduction**.

Most cells spend the majority of their life cycle in interphase, which is when they are doing their normal living activities. This makes sense—we spend most of our time living our life, and only a fraction, if any, in reproductive behavior. During interphase, cells go about their normal business, growing, maturing, and doing all the activities we have discussed and more. This is when the cell makes its contribution to the overall function of the whole organism.

During this phase, it also prepares for the reproduction to come. The DNA—your genetic material—replicates. That means it reproduces. Cells reproduce by dividing, but each daughter cell needs to have a complete set of all the DNA. So, during interphase, the complete set of DNA is copied in a process called **DNA replication.**

DNA replication occurs in the nucleus during interphase. During this process, all of your genetic code is reproduced exactly. The coiled

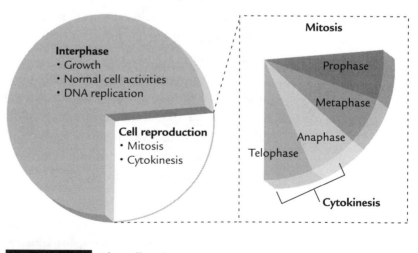

FIGURE 6.13   **The cell cycle.**

strands of DNA are held together by hydrogen bonds. When it is time to replicate, an enzyme breaks these bonds, allowing the strands to "unzip" (**Figure 6.14a**), which exposes their bases. Also in the nucleus are free nucleotides, which are the building blocks of your nucleic acids—each containing a sugar, a phosphate group, and a base. Once the DNA bases are exposed, the free nucleotides move in and search for partners. They are picky—adenine (A) only partners with thymine (T), and cytosine (C) always pairs with guanine (G). This type of match-up is called *complementary base pairing.*

When the DNA bases have the correct new nucleotides next to them, the sugars and phosphates of the free nucleotides are joined together by another enzyme. The result is an exact copy of the DNA strand that was previously there. Each original DNA strand is now mated with an exact copy of the other. Notice in **Figure 6.14c** that we now have twice as

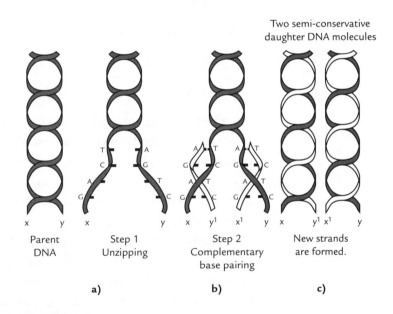

FIGURE 6.14 **DNA replication doubles our DNA before our cells divide. a)** In the first step, an enzyme unzips the strands of the original DNA, exposing their bases. **b)** Next, free nucleotides move in and align with the original bases through complementary base pairing (A with T, C with G). **c)** Finally, the free nucleotides are joined together to make a new strand of DNA that is identical to the original strand that it is replacing.

much DNA as when we started. That's the purpose of DNA replication—we double the DNA so that when the cell divides, each new cell gets a complete set of genetic instructions.

Once the DNA has replicated, cell reproduction may begin. Cell division, or reproduction, includes two processes:

- **mitosis**, which is nuclear division, and

- **cytokinesis**, which is cytoplasmic division.

The DNA is important to your cells—it controls all of their functions. Because of this, division of the nucleus is a separate process, called **mitosis**. This is a very precise event designed to ensure that the DNA is equally divided into the daughter cells.

Mitosis includes four phases:

- prophase

- metaphase

- anaphase, and

- telophase.

During interphase, the DNA is stretched out in thin strands called chromatin. During **prophase**, the first phase of mitosis, the chromatin condenses into rodlike structures—the chromosomes (**Table 6.2**). The nuclear membrane also disappears so that the chromosomes can move more freely. The next phase is **metaphase**. *Meta-* means *middle*, and during metaphase the chromosomes align very precisely, in duplicated pairs, along the midline, or equator, of the cell. This ensures that they will separate precisely. The next phase is **anaphase**. *Ana-* means *away*, and the duplicated chromosomes separate during this phase—one complete set goes to each side, or pole, of the cell. Movement of the chromosomes is directed by the centrioles and delicate structures that form from them, called spindle fibers, which pull the chromosomes in opposite directions. The final phase of mitosis is **telophase**, during which the chromosomes complete their journey to the opposite poles. This phase is like a reverse prophase. The nuclear membrane reappears and the chromosomes relax back into the stretched-out chromatin strands.

TABLE 6.2 **The stages of the cell cycle include interphase and cell reproduction.** Cell reproduction includes the four phases of mitosis during which the nuclear contents divide, and cytokinesis during which the remainder of the cell divides. Cytokinesis overlaps the latter phases of mitosis.

Picture	Stage	Events
	Interphase	Normal cell activities, growth, DNA replication. DNA is visible as thin strands called chromatin.
	Prophase	Chromatin condenses into rodlike structures called chromosomes that are clearly visible, and the nuclear membrane disappears.
	Metaphase	Chromosomes align very precisely along the midline.
	Anaphase	Chromosomes separate and are pulled to opposite poles of the cell. Cytokinesis begins once the chromosomes separate (visible here where the edge of the cell is just beginning to pinch in).
	Telophase	Chromosomes are in opposite poles and cytokinesis continues. This stage ends when the cells completely separate, forming two daughter cells.

TIME TO TRY

In the spaces provided, list the phases of mitosis in the correct order, then sketch what the cell would look like during that phase if it contained three pairs of chromosomes.

○ ○ ○ ○

_____   _____   _____   _____

Mitosis just divides the nucleus or, more specifically, the chromosomes. Once the chromosomes have carefully separated, the rest of the cell can be divided. This process, by which the cytoplasm divides, is called **cytokinesis**. It begins during anaphase of mitosis, after the chromosomes have separated, and ends at the end of telophase. At the end of cytokinesis, the original cell is gone and in its wake are two new daughter cells, almost identical and each with a complete set of DNA. These daughter cells are in interphase, and the whole cell cycle begins anew. Congratulations! Your cell just had babies.

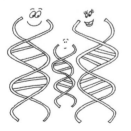 Cell reproduction includes mitosis, which is division of the nuclear contents, and cytokinesis, which is division of the cytoplasm. ■

Here's Looking at You, Kid! **Meiosis**

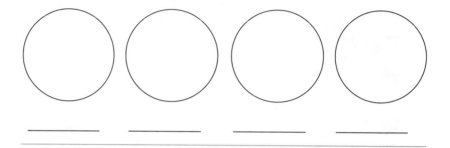

At the end of the normal cell life cycle, we get two daughter cells, each containing all 46 chromosomes—23 pairs. But when we make a baby, each parent contributes only half of the baby's genetic makeup. Our reproductive cells—sperm and ova—undergo a unique type of nuclear division called **meiosis,** which produces cells with only one chromosome from each pair. The division occurs in two cycles—meiosis I and meiosis II. The phases are the same as those of mitosis, but are also numbered according to the cycle. The key difference between mitosis and meiosis happens during metaphase. In mitosis, all 46 chromosomes line up individually, with their replicated copies, along the midline. When they separate during anaphase, the original separates from its copy, so each cell gets either an original or a copy of all 46 chromosomes.

In meiosis, though, chromosome pairs, not individual chromosomes, line up together at the midline, along with their replicated pairs. Then, in anaphase I, the pairs separate into different cells—each cell getting only one chromosome (and its copy) from each pair. At the end of telophase I, there are two daughter cells, each with only 23 chromosomes—one from each pair—still attached to its replicated copy. Because meiosis I cuts the chromosome count from 46 to 23, it is referred to as the *reduction division*. Meiosis II is similar to mitosis: Single chromosomes and their copies align along the midline, then separate during anaphase. Each daughter cell gets either an original or a copy of the 23 chromosomes. At the end of mitosis, there were only two daughter cells, but division occurs twice in meiosis, so there are four cells total. In males, that means four functional sperm, but in females only one of the daughter cells becomes a functioning ovum. Also, the daughter cells of mitosis are identical, but the cells produced by meiosis have unique combinations, allowing for genetic diversity and sometimes surprises when Baby arrives!

Final Stretch!

Now that you have finished reading this chapter, it is time to stretch your brain a bit and check how much you learned. For online tests, tutorials, animations, activities, web links, and an ebook, visit the *Get Ready for A&P* companion website.

RUNNING WORDS

At the end of each chapter, be sure you have learned the language. Here are the terms introduced in this chapter with which you should be familiar. Write them in a notebook or enter them into your computer. Define them in your own words, then go back through the chapter to check your meaning, correcting as needed. Also try to list examples when appropriate.

Fluid mosaic model
Cytosol
Nucleus
Chromatin
Chromosome
Nuclear envelope
Nuclear pore
Nucleoplasm
Ribosome
Protein synthesis
RNA (ribonucleic acid)
Messenger RNA (mRNA)
Ribosomal RNA (rRNA)
Nucleolus

Transfer RNA (tRNA)
Endoplasmic reticulum (ER)
Golgi apparatus
Vesicle
Mitochondrion
ATP
Lysosome
Peroxisome
Cytoskeleton
Microtubule
Microfilament
Centrosome
Centriole
Cilia
Flagellum
Brownian motion

Cytology
Cell
Cell Theory
Unicellular
Multicellular
Cell membrane
Cytoplasm
DNA
Prokaryote
Eukaryote
Nucleoid
Concentration gradient
Equilibrium
Simple diffusion
Facilitated diffusion
Osmosis
Solute
Solvent
Solution
Isotonic
Passive transport
Active transport
Pump
Exocytosis
Endocytosis
Phagocytosis
Pseudopodia

Plasmid
Organelle
Homeostasis
Plasma membrane
Phospholipid
Phosphate head
Hydrophilic
Fatty acid tail
Hydrophobic
Phospholipid bilayer
Selectively permeable
Phagosome
Receptor-mediated endocytosis
Pinocytosis
Endosome
Cell cycle
Interphase
Cell reproduction
DNA replication
Mitosis
Prophase
Metaphase
Anaphase
Telophase
Cytokinesis
Meiosis

WHAT DID YOU LEARN?

PART A: ANSWER THE FOLLOWING QUESTIONS.

1. List the five principles of the Cell Theory.

_____

_____

_____

_____

_____

2. What is the basic difference between prokaryotic cells and eukaryotic cells?

_____

_____

3. Describe the organization of the cell membrane. _____

_____

4. Design a concept map for the following terms: protein, DNA, nucleus, nucleolus, ribosome, rough ER, Golgi apparatus, and exocytosis.

5. Differentiate between passive and active movement processes. _____

_____

_____

6. In terms of movement processes, explain how making a cup of hot tea with a tea bag involves both osmosis and simple diffusion. _____

_____

_____

7. Assume you have limp carrot sticks in your refrigerator. To get them plump and crisp again, would you soak them in pure water or in a strong salt solution?

_____

8. Differentiate between phagocytosis and pinoctyosis. _____

_____

_____

9. List in order the four phases of mitosis:

_____

10. Explain the complete cell cycle. _____

_____

_____

_____

_____

PART B: FOR EACH OF THE FOLLOWING ITEMS, MATCH THE TERM WITH ITS DESCRIPTION.

1. mitochondrion _____

    a. Diffusion of water through a selectively permeable membrane.

2. lysosome _____

    b. Expelling materials out of the cell.

3. ribosome _____

    c. Site where ribosomes are made.

4. Golgi apparatus _____

    d. Process that uses energy to move molecules against their concentration gradients.

5. nucleolus _____

    e. Spontaneous movement of molecules down their concentration gradient.

6. pinocytosis _____

    f. "Garbage disposal" containing digestive enzymes.

7. exocytosis _____

    g. Site where proteins are made.

8. osmosis _____

    h. Site where cellular products are packaged.

9. active transport _____

    i. Cell drinking.

10. simple diffusion _____

    j. Site where ATP (energy) is made.

Answer Key

CHAPTER 1

Answers will vary from student to student.

CHAPTER 2

Part A
1. −32
2. 20,000
3. 3/4
4. 1/4
5. 1/4
6. 0.2
7. 720 breaths per hour
8. 400 cm
9. 30.48 cm
10. 120 lbs

Part B
1. 14
2. 4
3. 0.3; 30%
4. 20%
5. volume of a liquid
6. gram; meter (technically kilometer, but in practical use it is the meter); liter
7. "Normal" blood pressure is the average blood pressure.
8. 100°C
9. volume
10. 1000

CHAPTER 3

Part A
1. e
2. f
3. i
4. j
5. g
6. h
7. b
8. d
9. c
10. a

Part B
1. In the anatomical position, your palms are facing forward.
2. sagittal
3. d
4. b
5. coronal and sagittal
6. polycythemia; hepatitis
7. abnormal narrowing of the arteries; cartilage cell
8. pharynges; mitochondria; coxae
9. *Medial* is a comparative term meaning a designated object is closer to the midline; *median* means a structure is positioned exactly on the midline.
10. oblique

CHAPTER 4

Part A

1. organism, organ system, organ, tissue, cell, organelle, macromolecule, molecule, atom
2. The foot is a broadened surface with arches that distribute the weight evenly to all parts of the foot that contact the ground.
3. from the food we eat
4. If you consume too much energy (Calories), you gain weight; if you spend more energy than you consume, you lose weight; and if your caloric expenditure equals your caloric intake, your weight remains constant.
5. Homeostasis is the maintenance of a constant internal environment. It is important to maintain homeostasis because cells perform optimally in a balanced internal environment.
6. epithelial, connective, nervous, muscle

Part B

1. b, c
2. d, h
3. c, e
4. a
5. b, c, f
6. b
7. d
8. c, d, g

CHAPTER 5

Part A

	Potassium	Iodine	Oxygen	Neon
Chemical symbol	K	I	O	Ne
Atomic number	19	53	8	10
Atomic weight	39.10	126.90	16.00	20.18
Number of protons	19	53	8	10
Number of electrons	19	53	8	10
Number of neutrons	20	74	8	10
Number of electrons in outermost shell	1	7	6	8

Part B

1. protons and neutrons
2. electrons
3. carbon, hydrogen, oxygen, nitrogen
4. The row tells you how many shells of electrons that element has.
5. The column tells you how many electrons are in the outer shell of that element.

6. The calcium atom lost two electrons, giving it a $+2$ charge.
7. In ionic bonds, the electrons physically move from one atom to another so that the atoms involved lose or gain electrons, and the opposite charges of the resulting ions draw the atoms together. In covalent bonding, the atoms share electrons.
8. structural
9. decomposition (catabolic hydrolysis)
10. synthetic (anabolic, dehydration synthesis)
11. carbon and hydrogen
12. $C_6H_{12}O_6$ O
    CO_2 I
    CH_4 O
    CO   I
    HCl   I
    H_2O I
13. monosaccharides for carbohydrates; amino acids for proteins
14. protein
15. ATP (adenosine triphosphate)

CHAPTER 6

Part A

1. All organisms are composed of one or more cells; cells are the basic structural and functional units of life; all vital functions of an organism occur within cells; all cells come from preexisting cells; and cells contain hereditary information that regulates cell functions and is passed from generation to generation.
2. Prokaryotes have no nucleus and lack internal membranes.
3. The cell membrane is a phospholipid bilayer with protein channels periodically passing through it and other molecules such as cholesterol and carbohydrates "floating" in it.
4. Concept maps will vary from student to student, but these relationships should be included:
   - DNA contains the instructions for how to build proteins and is located in the nucleus.
   - The nucleolus is located inside the nucleus and makes ribosomes.
   - Rough endoplasmic reticulum gets its appearance from the presence of ribosomes.
   - Proteins are made at ribosomes, so rough ER makes proteins.
   - Proteins from rough ER move to the Golgi apparatus for processing and packaging, then are expelled from the cell by exocytosis.
5. In passive movement processes, substances move along their concentration gradients from high concentration to low concentration. In active movement processes (active transport), energy is used to move substances against their concentration gradients (from low concentration to high).
6. When a tea bag is placed in hot water, the water moves into the tea bag by osmosis and dissolves the tea, then the tea molecules diffuse out of the tea leaves into the water.

7. Water would move into the carrot, plumping the cells and making them rigid. (Salt would draw more water out, causing them to shrivel even more.)
8. In phagocytosis, the cell surrounds a solid, with the cell membrane extending out to surround the object, then drawing it inward. In pinocytosis, the cell membrane forms an inward pouch that surrounds a droplet of liquid.
9. prophase, metaphase, anaphase, and telophase
10. The complete cell cycle includes all events that occur in the life of a cell, specifically interphase and cell reproduction. Interphase is when the cell carries out its normal functions, and the DNA replicates in preparation for reproduction. Cell reproduction includes the four phases of mitosis that divide the nucleus, and cytokinesis that divides the cytoplasm.

Part B

1. j
2. f
3. g
4. h
5. c
6. i
7. b
8. a
9. d
10. e

Index

Photo and Illustration Credits

Unless noted, all chapter opener art and cartoon spot art was created by Kevin Opstedal.

CHAPTER 1

Table 1.2: All photos from PhotoDisc.

Figures 1.1, 1.2, 1.3, 1.4: Seventeenth Street Studios.

Figure 1.5: Tom Stewart/CORBIS.

CHAPTER 2

Figures 2.1, 2.2, 2.3, 2.4, 2.5: Seventeenth Street Studios.

Figure 2.6: Benjamin Cummings Publishing, Pearson Education.

Figure 2.7a: Ohaus Corporation.

Figure 2.7b: Seventeenth Street Studios.

Figure 2.8: Richard Megna/ Fundamental Photographs.

Figures 2.9, 2.10, 2.11: Seventeenth Street Studios.

CHAPTER 3

Figures 3.1, 3.2: Seventeenth Street Studios.

Figure 3.3; hip implant: National Institutes of Health. Art by Seventeenth Street Studios.

Figure 3.4: Victoria & Albert Museum, London/Art Resource, NY.

Figures 3.5, 3.6: Kevin Opstedal.

Figure 3.7: Seventeenth Street Studios.

CHAPTER 4

Figure 4.1; tissue: Benjamin Cummings Publishing, Pearson Education.

Figure 4.1; cell: Ed Reschke/Peter Arnold.

Figure 4.1; organelle: Don W. Fawcett/Visuals Unlimited.

Figure 4.1; art: Seventeenth Street Studios.

Figure 4.2: Seventeenth Street Studios.

Figure 4.3a, b: Frederick Skvara/Visuals Unlimited.

Figure 4.3c, d, e: Ralph T. Hutchings.

Figure 4.4a: P. Motta/Photo Researchers.

Figure 4.4b: Steve Gschmeissner/SPL/Photo Researchers.

Figure 4.5a, b, c: Benjamin Cummings Publishing, Pearson Education.

Figure 4.5d: Carolina Biological/Visuals Unlimited.

Figures 4.5e, f: Benjamin Cummings Publishing, Pearson Education.

Figure 4.6: David L. Bassett

Figure 4.7: Adapted from Marieb, *Essentials of Human Anatomy & Physiology*, 8e, F1.2, © Benjamin Cummings, 2006.

CHAPTER 5

Figure 5.1; tissue: Benjamin Cummings Publishing, Pearson Education.

Figure 5.1; cell: Ed Reschke/Peter Arnold.

Figure 5.1; organelle: Don W. Fawcett/Visuals Unlimited.

Figure 5.1; art: Seventeenth Street Studios.

Figures 5.2, 5.3, 5.4, Time to Try, p. 172, 5.5, 5.6, 5.7, 5.8, 5.9, 5.10: Seventeenth Street Studios.

Figure 5.8, photo: Lori Garrett.

Figure 5.12: Louis Pepin.

CHAPTER 6

Figure 6.1; tissue: Benjamin Cummings Publishing, Pearson Education.

Figure 6.1; cell: Ed Reschke/Peter Arnold.

Figure 6.1; organelle: Don W. Fawcett/Visuals Unlimited.

Figure 6.1, art: Seventeenth Street Studios.

Figure 6.2a: Dennis Kunkel/Visuals Unlimited.

Figure 6.2b: David McCarthy/Photo Researchers.

Figure 6.2c: David M. Phillips/Photo Researchers.

Figure 6.4: Seventeenth Street Studios.

Figure 6.5: Adapted from Marieb, *Study Guide* for *Human Anatomy & Physiology,* 6e, F3.1, © Benjamin Cummings, 2004.

Figure 6.6: Seventeenth Street Studios.

Figure 6.7: D.W. Fawcett/Photo Researchers/Adapted from Bauman, *Microbiology*, F3.37, © Benjamin Cummings, 2004.

Figures 6.8, 6.9, 6.10: Seventeenth Street Studios.

Figure 6.11a, b, c: M. Sheetz, R. Painter, and S. Singer, *Journal of Cell Biology* 1976. 70: 193.

Figure 6.12: Birgit H. Satir, Dept. of Anatomy and Structural Biology, Albert Einstein College of Medicine.

Figure 6.13: Seventeenth Street Studios.

Table 6.2: All photos by Ed Reschke.